CREEPY INSECTS & TINY MONSTERS

A MIND-BLOWN™ SERIES FUN BOOK

1000+ Fun & Weird Facts, Myths, Legends, Stories +

THE ULTIMATE FUN, WEIRD AND WONDERFUL BOOK ON THINGS THAT THRILL, SCARE AND MAKE US SAY "EWW"

I0101298

MIND-BLOWN

ISBN: 979-8-9938493-4-8
MIND-BLOWN™ ENTERTAINMENT & PUBLISHING TEAM
  IN ASSOCIATION WITH WG COAKLEY PUBLISHING LLC
  ILLUSTRATIONS: LUMEN GARY, ALDEN SKETCH, KAREN SATO
  RESEARCHERS: B.G SCOUT, WILDER FACTUM, IAN PATEL, JO VAN
IMAGE ATTRIBUTION TO: FREEPIK – FLATICON.COM, ADOBE, NASA, USGS, NSF

PART OF THE MIND-BLOWN™ FUN BOOK SERIES
FIRST EDITION
PRINTED IN THE UNITED STATES

The Official Stamp of Awesome Weirdness

This is **NOT** another sleepy bug book.

This is a **MIND-BLOWN**^{tm} book.

Inside, you'll find:

Fun & Weird Facts - quick-hit exoskeleton mind-benders

Myths – Busted - backyard myths crushed by real science

Legends - eerie, maybe-true tales from the undergrowth

Did You Know? - deeper insights into the micro-verse

Story Moments - short scenes that shrink you to bug-size

Fun Quizzes - fast challenges to test your knowledge

Jokes & Comics - Mind-Blown™ weird and wiggly

Mini-Games & Challenges - quick, fun activities

Bonus Extras , Resources with QR Codes

Insects and Tiny Monsters.
Real science style images.

Warning: Bugs can be a bit scary. Some insect images and info may not be deemed appropriate for very young children. Adult review and decision recommended.

CONTENTS

Bug Smiles

A man had a pet centipede. He said, "centipede, go get the paper and make it snappy!"

A half an hour later the man went outside and said, "I thought I told you to get the paper a half an hour ago!"

"Well, I had to put on my shoes," said the centipede.

CHAPTER 1

1-Masters of Disguise
Insects that Cheat Reality

Shape-shifting, mimicry, camouflage, deception

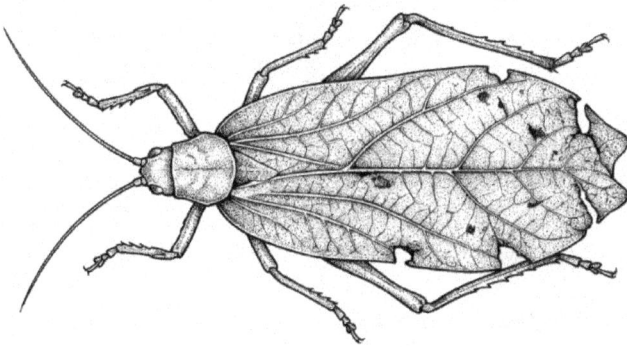

Dead Leaf Katydid
Typophyllum spp.

COMMON LOCATION: Central and South America
Resembles a dead leaf to avoid predators; remains motionless during the day.

FUN & WEIRD FACTS

Some Assassin Bugs wear the dead bodies of their enemies. After draining the fluids from their prey, **Assassin Bugs** (predatory insects found in diverse habitats worldwide) glue the empty exoskeletons onto their backs using sticky secretions. This gruesome backpack acts as camouflage to break up their outline and confusing larger predators. A pile of ants walking on a leaf looks much less tasty than a juicy bug.

The Orchid Mantis is a deadly trap hidden in beauty. **The Orchid Mantis** (a flower-mimicking predator found in Southeast Asian rainforests) resembles a pink and white orchid blossom so perfectly that hungry insects fly straight into its claws. It actually attracts more pollinators than the real flowers nearby.

Spicebush Swallowtail Caterpillars look like cartoon snakes. To scare away birds, **Spicebush Swallowtail Caterpillars** (larval insects found in North American woodlands) display huge fake eyespots on their heads and puff up their bodies.

Giant Atlas Moths have snakes painted on their wings. The tips of the wings on **Giant Atlas Moths** (massive moths found in Asian forests) look exactly like the heads of cobras, complete with eyes and scales. When the moth falls to the ground and flutters, it looks like a writhing snake to scare off attackers.

Green Lacewing Larvae build suits of trash. **Green Lacewing Larvae** (voracious predators found in gardens) impale debris and leftover food skins onto bristles on their backs. This "trash packet" hides them from birds and allows them to sneak up on aphids without being recognized.

Bagworm Moths live inside mobile log cabins. **Bagworm Moths** (caterpillars found on trees worldwide) construct tough, protective cases out of silk and twigs that they carry with them everywhere. They never leave this home until they mature, blending in perfectly with tree branches.

Spittlebugs hide inside their own bubbles. **Spittlebugs** (sap-sucking insects found in meadows and gardens) pump air into a liquid they secrete to create a foamy fortress that looks like spit on a plant stem. This wet blanket hides **Spittlebugs** from view and protects them from drying out or being eaten. It is a gross but effective way to stay safe.

Thorn Bugs look painfully sharp to eat. **Thorn Bugs** (treehoppers found in tropical regions) have hardened, pointed bodies that look exactly like the jagged thorns of the plants they live on. Birds skip over them, assuming they are just a sharp piece of the plant.

Thorn Bug
Membracidae family (adult)

COMMON LOCATION: Tropical and subtropical forests, especially Central and South America.

DESCRIPTION: The thorn bug looks less like an insect and more like a weaponized plant spine. Its hardened, sharply pointed body perfectly mimics the jagged thorns of the stems it lives on, making predators hesitate—or pass it by entirely. Birds scanning for soft prey often ignore thorn bugs completely, mistaking them for nothing more than a dangerous piece of the plant itself.

Bird Dung Spiders look exactly like fresh bird droppings. **Bird Dung Spiders** (crab spiders found in forests) curl up and lie motionless on leaves to mimic white and black waste. Predators usually have no desire to eat poop, so the spider stays safe.

Walking Leaves mimic the damage of real plants. **Walking Leaves** (leaf insects found in South Asia and Australia) often have brown spots or jagged edges on their bodies that look like rot or bite marks. This imperfection makes the disguise convincing because real leaves are rarely perfect.

The Dead Leaf Butterfly vanishes when it lands. With its wings closed, the **Dead Leaf Butterfly** (a nymphalid butterfly found in tropical Asia) displays a pattern of veins, mold spots, and dried brown colors. It looks so much like a fallen leaf that it is nearly impossible to spot on the forest floor.

Dead Leaf Butterfly
Kallima inachus

COMMON LOCATION: South and Southeast Asia.

DESCRIPTION: When its wings are closed, this butterfly looks almost exactly like a dry, dead leaf, complete with veins and bite marks, allowing it to vanish in plain sight from predators.

Some Stick Insects sway to the rhythm of the wind. **Stick Insects** (camouflaged herbivores found in forests) rock back and forth when on a branch to mimic the movement of twigs in a breeze. This behavior prevents their movement from catching the eye of a hunter.

Clearwing Moths have transparent windows in their wings. **Clearwing Moths** (day-flying moths found in the Northern Hemisphere) lack the colored scales that most moths have, making their wings see-through like a wasp or fly. This optical illusion tricks predators into thinking they are dangerous stinging insects.

Cellar Spiders vibrate wildly to become invisible. When threatened, **Cellar Spiders** (common household arachnids found in basements) shake their webs so fast they turn into a confusing blur. This dizzying motion makes it impossible for a predator to aim an attack.

Silverfish move like liquid mercury across the floor. Silverfish (wingless insects found in damp bathrooms) wiggle their bodies in a fluid motion that mimics a swimming fish. This erratic movement makes it incredibly difficult for humans or spiders to catch Silverfish.

Carpet Beetle Larvae disguise themselves as dust bunnies. Carpet Beetle Larvae (small pests found in rugs and closets) are covered in bristly hairs that make them look exactly like a piece of fuzzy lint. Most people mistake them for harmless dirt. You might vacuum up Carpet Beetle Larvae without ever realizing they are alive.

Cockroaches can flatten their bodies to vanish into thin air. Cockroaches (scavenging insects found in kitchens) can splay their legs to squeeze into cracks as thin as a quarter. This ability allows Cockroaches to disappear into walls instantly, as if by magic.

Pill Bugs roll into perfect, armored stones. Pill Bugs (crustaceans found in gardens and basements) curl into a tight ball when touched, hiding their soft legs inside a hard shell. To a predator, a Pill Bug looks just like an inedible pebble.

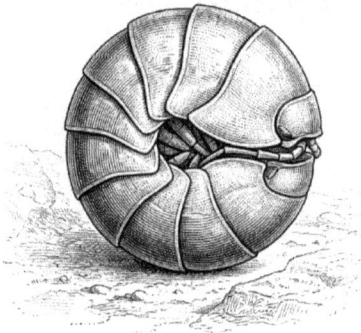

Pill Bug
Armadillidiidae family (adult)

COMMON LOCATION:
Gardens, forests, and damp indoor spaces worldwide.

DESCRIPTION:
Pill bugs turn themselves into perfect living stones. When touched, they curl into a tight armored ball, sealing their soft legs and underside inside a hardened shell. To a predator, a rolled pill bug doesn't look alive at all—just a small, inedible pebble that isn't worth biting.

House Centipedes detach their own legs to trick enemies. If trapped, House Centipedes (fast-moving arthropods found in homes) will drop a leg that continues to twitch and wiggle on the ground. This distracting decoy keeps the predator busy while the House Centipede escapes on its remaining legs.

Stink Bugs use an invisible chemical wall. Stink Bugs (shield-shaped insects found in homes) do not need to run fast because they carry a terrible smell. The threat of this odor acts as a "do not touch" sign that most predators respect.

Ghost Ants look like floating dark spots. Ghost Ants (tiny household pests found in kitchens) have pale, translucent legs and abdomens that are nearly invisible to the naked eye. Only their dark heads are easily seen, confusing homeowners who see tiny specks moving on the counter.

House Flies see the world in slow motion. House Flies (common flying pests found globally) process visual information seven times faster than humans do, allowing them to dodge swatters with ease.

House Fly
Musca domestica

COMMON LOCATION:
Worldwide.

DESCRIPTION:
In slow motion, a house fly reveals rapid wingbeats, sudden direction changes, and precise body control, allowing it to take off, turn, and evade danger faster than the human eye can track.

Common House Spiders are excellent actors. When cornered, Common House Spiders (web-building arachnids found indoors) will curl their legs and remain perfectly still. By playing dead, Common House Spiders trick humans into sweeping them up or walking away.

Hoverflies perform a dangerous act called Batesian mimicry. Hoverflies (pollinators found in gardens worldwide) are harmless and cannot sting, but they have yellow and black stripes just like bees/wasps. Birds avoid **Hoverflies** because they remember the pain of a bee sting.

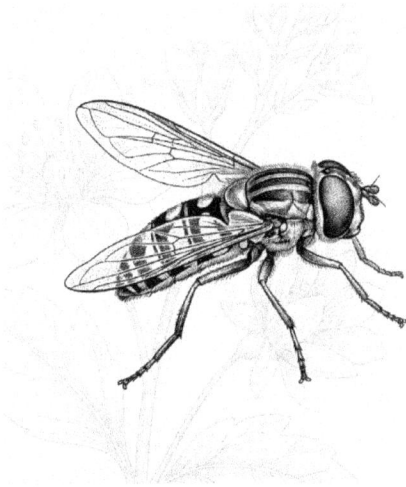

Hoverfly

Syrphidae family (adult)

COMMON LOCATION: Gardens, meadows, and farmland worldwide.

DESCRIPTION: Hoverflies pull off one of nature's boldest bluffs. Though completely harmless and unable to sting, their yellow-and-black striping closely mimics bees and wasps. Birds that have learned the pain of a real sting often avoid hoverflies entirely, fooled by appearance alone.

Golden Tortoise Beetles can change color to liquid gold. When threatened, Golden Tortoise Beetles (small leaf beetles found in the Americas) alter the fluid flow in their shell to turn a brilliant, metallic gold. This sudden flash reflects light and confuses predators long enough for the beetle to escape.

Lanternflies use fake eyes to startle attackers. When threatened, Lanternflies (planthoppers found in tropical forests) spread their wings to reveal bright red and black eye-spots that mimic a much larger animal.

Wrap-around Spiders flatten themselves to become part of the branch. Wrap-around Spiders (orb-weavers found in Australia) have concave bellies that allow them to hug a twig tightly, disappearing into the bark. During the day, they look like a natural bump on the wood.

Ant-mimicking Spiders pretend to be insects. Ant-mimicking Spiders (jumping spiders found globally) walk on six legs and wave their front two legs in the air to look like antennae. Since ants taste bad and bite back, predators leave the spider alone.

Viceroy Butterflies copy the poisonous Monarch. Viceroy Butterflies (orange and black butterflies found in North America) evolved to look nearly identical to the toxic Monarch butterfly. Birds learn to avoid the bright orange warning colors, keeping the tasty Viceroy Butterfly safe.

Viceroy Butterfly
Limenitis archippus (adult)

COMMON LOCATION: Meadows, wetlands, and forest edges across North America.

DESCRIPTION: Viceroy butterflies survive by borrowing danger they don't actually possess. Though edible and harmless, they evolved to closely resemble the toxic Monarch butterfly, copying its bolder **bold orange-and-black warning colors**. Birds that remember the sickness caused by Monarchs often avoid Viceroys as well, mistaking mimicry for poison.

Underwing Moths use flash coloration to startle enemies. Underwing Moths (nocturnal insects found in temperate zones) have dull gray upper wings, but they flash bright red or orange lower wings when disturbed. This sudden burst of color shocks the predator, giving the moth a split second to fly away.

Puss Moth Caterpillars wear a terrifying mask. Puss Moth Caterpillars (larvae found in European woodlands) have a head that looks like a scary red face with black eyes to frighten birds. If that fails, they can spray acid from their tails.

Lobster Moth Caterpillars look like alien monsters. Lobster Moth Caterpillars (insects found in Palearctic forests) have bizarre, lumpy shapes and long legs that make them look more like wood debris or a crushed spider than a juicy caterpillar.

Ladybird Mimic Spiders can pretend to be beetles. Ladybird Mimic Spiders (arachnids found in Asia) have bright orange bodies with black spots to mimic the toxic Ladybug. This warning color tells birds that the spider will taste terrible.

Stick Insect eggs look like plant seeds. Stick Insects (phasmids found worldwide) drop hard, round eggs that resemble seeds to hide them on the forest floor. Ants often carry these eggs underground, protecting them until they hatch.

Water Scorpions look like sunken twigs. Water Scorpions (aquatic insects found in ponds) hang motionless in the water with a breathing tube that looks like a dead plant stem. Small fish swim right past them, unaware of the danger until the Water Scorpion strikes.

Elephant Hawk-moth Caterpillars, mimic a snake's head when threatened. Elephant Hawk-moth Caterpillars (larvae found in Europe) retract their real head and inflate the front of their body, revealing large false eyes that resemble a reptile's stare. The sudden shape change can startle birds and small mammals into backing away. For a brief moment, a harmless leaf-eating grub convincingly passes as a dangerous viper.

Bug Smiles

A fly was sitting on a piece of poop.
Another fly said, "Hey, want to go grab lunch?"

The first fly said, "No thanks. I just ate."

Moss Mantises mimic the soft texture of forest growth. Moss Mantises (small predators found in Africa) have bumps and frills on their exoskeleton that look like patches of moss. This allows Moss Mantises to hide in plain sight while waiting for prey.

Moss Mantis
Haania spp.

COMMON LOCATION: Southeast Asia.

DESCRIPTION: This mantis mimics the soft, uneven texture of moss-covered bark and forest growth, using rough body surfaces and muted coloration to disappear completely against trees and rocks.

Buff-tip Moths look exactly like a broken birch twig. Buff-tip Moths (nocturnal moths found in Europe) roll their wings around their bodies to form a cylinder with a "broken wood" pattern at the tip. Birds land right next to them without realizing a meal is present.

Cuckoo Wasps roll into an armored ball. When under attack, Cuckoo Wasps (parasitic wasps found globally) curl into a tight sphere, exposing only their hard, jeweled exoskeleton. This protects their soft belly and makes them nearly impossible to bite.

Ironclad Beetles are too tough to chew. **Ironclad Beetles** (desert beetles found in North America) have an exoskeleton so hard that you cannot push a pin through it. Predators quickly learn that Ironclad Beetles are a menu item that will break their teeth.

Leaf-footed Bugs have leaves growing on their legs. Leaf-footed Bugs (sap-sucking insects found in warm climates) have flat, leaf-like expansions on their hind legs that break up their outline. This helps them blend in with dense foliage.

Owl Butterflies have giant eyes on their wings. Owl Butterflies (large butterflies found in rainforests) have massive spots that look like the eyes of an owl, a natural predator of small lizards and birds. Seeing these eyes makes attackers freeze in fear.

Jewel Beetles use mirrors to hide. Jewel Beetles (metallic wood-boring beetles found globally) have iridescent shells that reflect the green and gold colors of the forest. This dazzling reflection breaks up their shape in dappled sunlight, making **Jewel Beetles** hard to spot.

Jewel Beetle
Buprestidae

COMMON LOCATION: Forests and woodlands worldwide.

DESCRIPTION: Jewel beetles have mirror-like, iridescent shells that reflect green and gold forest light, breaking up their outline in dappled sunlight and making them surprisingly difficult to spot.

Scorpion flies have a fake stinger. Scorpionflies (flying insects found in damp woodlands) have a tail that curls up like a scorpion's stinger, but it is actually harmless reproductive equipment.

Robber Flies mimic fuzzy bumblebees. Robber Flies (aggressive predatory flies found worldwide) evolved hairy bodies and black-and-yellow coloring to look like stinging bees. This disguise lets Robber Flies hang out near flowers to ambush real bees.

Young Stick Insects mimic venomous scorpions. Stick Insects (specifically the Giant Prickly Stick Insect found in Australia) curl their tails over their backs when they are babies to look like scorpions. This trick keeps birds away until the **Stick Insects**

grow large enough to rely on camouflage.

Ghost Mantises look like withered, brown leaves. Ghost Mantises (small predators found in Africa) have a twisted body shape and dark colors that resemble dry, dead foliage. This gloomy disguise is perfect for ambush hunting in dry seasons.

Mantispid Flies are a mashup of two dangerous bugs (predatory insects found in subtropics) look like a cross between a praying mantis and a wasp. This double threat confuses predators who aren't sure which weapon to fear.

Ambush Bugs have irregular, jagged bodies. Ambush Bugs (predatory insects found on flowers) are shaped like broken plant matter to hide inside colorful blooms. They wait perfectly still for a bee to land, then strike.

Grasshoppers in burnt forests turn black. Grasshoppers (insects found in fire-prone regions) can develop dark coloration to match the charred wood of a forest after a fire. This adaptation allows Grasshoppers to survive in an ash-covered landscape.

Fire-Melanistic Grasshopper
Melanoplus spp.

COMMON LOCATION: Fire-prone grasslands and forests worldwide.

DESCRIPTION: After wildfires, some grasshopper populations develop darker coloration that blends into ash and charred wood, helping them avoid predators in newly burned landscapes.

FUN BUG GAMES
FAMILY OR PERSONAL FUN

Play online:

The "**Big Bug Memory Game**" is a popular online matching game from **National Geographic Kids**, where you find pairs of insects, testing your memory across different difficulty levels (easy, medium, hard) with various bugs like bees, beetles, and spiders.

 You can play it directly on their website by searching for "Nat Geo Kids bug memory game" or find similar insect-themed matching games on ESL sites or apps for learning.

 Seek by iNaturalist lets you discover and identify real insects using your phone's camera—turning the outdoors into a real-life bug hunt. **This app turns your backyard, park, or sidewalk into a real-life bug hunt.**

☑ **Created by iNaturalist** (California Academy of Sciences + National Geographic)

Stick Insect
Phasmatodea

COMMON LOCATION: Forests and woodlands worldwide, especially in tropical regions.

DESCRIPTION: Stick insects have long, twig-like bodies that match the shape, color, and even texture of branches, allowing them to **remain motionless** and **nearly invisible** among plants for hours at a time.

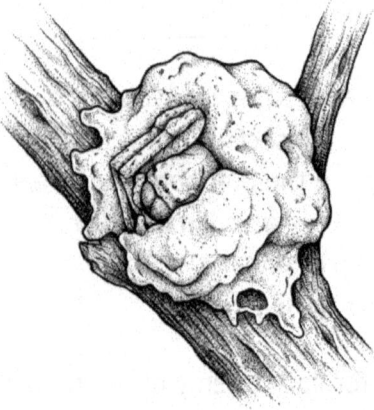

Bird-Dropping Spider
Celaenia spp.

COMMON LOCATION:
Australia and parts of Asia.

DESCRIPTION:
This spider disguises itself as a splatter of bird droppings, using pale coloring, lumpy shape, and stillness to avoid predators while also luring insects that mistake it for harmless debris.

MYTHS - BUSTED

Many people believe camouflage insects can change color instantly, just like chameleons. In reality, almost no insects are capable of rapid color shifts. Most changes occur slowly across molts or environmental adjustments, relying more on posture and texture than on color.

A common myth claims leaf insects freeze completely when spotted. In truth, most species maintain subtle swaying motions that mimic natural leaf movement. Freezing would make them stand out more, not less.

Some people think flower mantises mimic blossoms mainly to avoid predators. Their floral disguise is actually an offensive adaptation-they lure prey such as bees and butterflies that mistake them for real flowers.

There's a myth that stick insects stay motionless for hours. Many species reposition frequently, timing micro-movements with the rhythm of surrounding leaves to maintain their camouflage.

People often assume leaf-mimicking insects are fragile because they resemble delicate foliage. In reality, many leaf insects have surprisingly sturdy exoskeletons designed to support their broad wing structures.

Some believe thorn-mimicking insects are poisonous. Most thorn mimics are harmless; predators avoid them because they resemble real thorns, not because they pose a toxic threat.

Bug Smiles

Q: What is totally funny and makes dogs itch?
A: The Flea Stooges!

‾‾‾‾‾‾-

Q: What do you call a rabbit with beetles all over it?
A: Bugs Bunny.

‾‾‾‾‾‾-

Q: Why was the ant so confused?
A: Because all his uncles were "ants"!

‾‾‾‾‾‾-

Q: What creature is smarter than a talking parrot?
A: A spelling bee!

‾‾‾‾‾‾-

Q: What is on the ground and a hundred feet in the air?
A: A centipede on its back!

LEGENDS

LEGENDS

In northern Borneo (Malaysia) , Dayak hunters long told of an insect called the "Walking Shadow," said to vanish the moment someone tried to look directly at it. British naturalists in the late 19th century recorded notes describing insect outlines that seemed to fade between sketches. Locals claimed the creature only moved with the forest's rhythm, disappearing when observed too intently. Modern biologists attribute these sightings to extreme crypsis, where phasmids synchronize their posture with shifting light, creating the illusion of vanishing mid-glance.

In Madagascar's eastern rainforests, Betsimisaraka villagers speak of the "Leaf That Watches," a silent insect believed to follow travelers with its gaze. Early French naturalists documented mantises with wing markings resembling large, staring eyes. Villagers insisted the insects appeared only on certain paths during fog. Scientists now know these eye-spots function as predator deterrents, but the eerily human-like symmetry kept the legend alive for generations.

Among the Karen people of northern Thailand, elders describe insects that "borrow the daylight," appearing only when the sun dips low. Explorers in the early 1900s reported phasmids aligned so precisely with long dusk shadows that their bodies appeared and disappeared as the light shifted. The effect-shadow-matching-is a rare camouflage strategy that creates the impression of creatures phasing in and out of existence.

In Papua New Guinea's Huli highlands, stories tell of the "Tree Whisperer," an insect that moves only when no one is watching. Ethnographers documented accounts of phasmids whose swaying matched the micro-movements of eucalyptus bark in the wind. Hunters claimed the insects changed position only between blinks, an observation consistent with motion-timed camouflage-where insects move exclusively during ambient movement, hiding their transitions.

In India's Western Ghats, villagers warned of the "Cradle Leaf," a mimic said to hang like a seed pod and sway even without wind. British survey teams in the 1930s found leaf insects suspended in perfect pod-shapes, their posture tightened into arched curves. Even slight weight shifts caused eerie pendulum swings. Biologists later identified this as an adaptive behavior to mimic monsoon-season seed dispersal, though locals believed the motion foretold approaching storms.

In Costa Rica's cloud forests, guides tell of the "Lost Vein," a leaf insect that disappears as soon as someone tries to trace its wing patterns. Researchers studying wing venation observed that certain species redirect attention through a visual trick: when focused on the veins, the peripheral outline softens and seems to vanish. Visitors believed the insect was slipping between realities; scientists recognized it as one of nature's most effective attention-misdirection camouflages.

In Japan's mountain villages, folklore speaks of the "Kage-Mushi," or shadow bug, said to shift between twig-like and bone-white forms under moonlight. Entomologists found phasmids with wax-coated exoskeletons that reflect moonlight, giving them a pale, skeletal look at night while appearing brown and twig-like by day. Because the transformation depends entirely on lighting angle, villagers believed they were seeing two different creatures inhabiting the same forest spirit.

DID YOU KNOW?

Did you know? Some **stick insects** regenerate lost limbs with altered proportions, helping them blend in even better as they age. Their bodies grow asymmetrically to match broken twigs or damaged leaves around them, creating more convincing camouflage. This regeneration ability also helps them survive predator attacks by sacrificing a limb and later regrowing it during molting.

Did you know? Leaf insects can match not just color but translucency, causing their wing edges to glow like real leaves under sunlight. The veins in their wings even align like leaf veins, strengthening the illusion for both predators and prey. Scientists have discovered that this mimicry can change slightly with humidity and light levels, fine-tuning their invisibility in different forest conditions.

Did you know? Some **katydids** send micro-vibrations through leaves to detect predators without exposing themselves to movement-sensing birds. These subtle vibrations act like sonar, letting them "feel" the surroundings through tremors in the leaf surface. Males also use similar vibrations to communicate with females, ensuring they stay hidden while finding mates.

Did you know? Certain **mantises** shift posture rather than color, imitating broken stems or dying leaves with impressive precision. They sway gently to mimic plants moving in the breeze, helping them approach prey unseen. This behavioral mimicry is often more important than color change because many predators spot motion faster than shape.

🐛 **Did you know? Bark insects** align perfectly with cracks matching their width, eliminating the shadow lines predators use for detection. Their flat bodies and muted brown tones let them vanish against tree trunks during the day. Some species even accumulate lichen and dust on their backs over time, further blending into the bark texture.

🐛 **Did you know?** Many insects move only when the wind moves surrounding plants, timing their motion with the environment to stay hidden. This "camouflage synchronization" prevents predators from detecting motion contrast - one of the main triggers for attack. Observations show that grasshoppers, stick insects, and leafhoppers all pause instinctively the moment the air becomes still.

Leaf Insect
Phyllium spp.

COMMON LOCATION: Southeast Asia and nearby regions.
Disguises itself as a leaf to avoid being seen by predators.

STORY MOMENT

The Leaf That Shifted Twice

During a late-afternoon survey in the **Cameron Highlands of Malaysia, entomologist Rina Solano** paused when she noticed what looked like two overlapping leaves pressed against a mossy branch. The forest was warm and hazy, the kind of light that turns every detail into a quiet shimmer. She leaned closer, thinking she had found a particularly symmetrical leaf cluster- until the upper "leaf" shifted by a fraction of a degree. It wasn't wind. It wasn't coincidence. It was intention.

As she steadied her flashlight, the second "leaf" adjusted its midrib, revealing a subtle but unmistakable living posture. Both insects realigned themselves with the fading sunlight, their bodies flattening into perfect silhouettes that matched the branch's dappled pattern. Even as she watched, their edges softened and blended, dissolving into the scenery in real time. Rina wasn't witnessing camouflage-she was witnessing participation in the landscape.

That night, reviewing her notes, she wrote a single line that stayed with her for years: "Camouflage isn't hiding. It's becoming part of the story the forest is already telling."

Bug Smiles

A lonely old man decides to get a pet centipede..

He takes the pet centipede home and sets up a cage for him.
The next morning, the man goes up to the cage and asks the
centipede, "Hey, would you like to go out to breakfast with me?"
The centipede does not respond.

Lunch comes around and the man again goes to the cage and
asks, "Would you like to go to lunch with me?"
The centipede still does not respond and the man walks away sad.

Dinner comes around and again, the man goes to the cage and
asks, "Hey would you please like to go to dinner with me?"

To which the centipede responds, "I heard you the first time! I'm
putting on my shoes!"

FUN QUIZ

1. Which insect creates tiny holes to mimic damaged leaves?

A) Bark beetle B) Leaf insect C) Dragonfly D) Lacewing

2. What movement pattern helps stick insects avoid detection?
A) Hopping B) Zig-zag sprinting C) Gentle swaying D) Freezing

3. Which spectrum allows birds to see hidden patterns on moth wings? A) Infrared B) Ultraviolet C) X-ray D) Red

4. True or False: Camouflaged insects rely mostly on rapid color changes.

5. True or False: Leaf insects can appear to glow at the edges when sunlight passes through their wings.

6. Which insect group often mimics lichen-covered bark?

A) Treehoppers B) Grasshoppers C) Mosquitoes D) Cicadas

7. What communication method do some katydids use to stay hidden?

A) Loud clicking B) Leaf vibrations C) Eye flashes D) Color shifting

8. True or False: Flower mantises mimic blossoms mainly to avoid predators

QUIZ ANSWERS

1: B) Leaf insect

2: C) Gentle swaying

3: B) Ultraviolet

4: False

5: True

6: B) Grasshoppers

7: B) Leaf vibrations

8: False

Bug Smiles

Q: What do frogs order when they go to a restaurant?
A: French Flies.

———-

Q: What goes 99 thump,99 thump,99 thump?
A: A centipede with a wooden leg.

CHAPTER 2

2-Poisoners, Venom-Shooters & Chemical Wizards

Chemistry, toxins, sprays, acids, explosive reactions

Bombardier Beetle Brachinus spp.
COMMON LOCATION: Worldwide Produces explosive
chemical sprays to deter predators.

FUN & WEIRD FACTS

MIND-BLOWN MOMENT

Some bombardier beetles fire boiling chemicals from their rear end. They mix hydrogen peroxide and hydroquinone inside a reaction chamber and detonate it, producing a jet of scalding, toxic spray that reaches predators mid-attack.

Assassin bugs liquefy their prey before drinking them. They inject a chemical cocktail of digestive enzymes that turns the insides of insects into a soup, which they then suck out like a milkshake.

Fire ants spray venom onto their own bodies to weaponize themselves. They coat their exoskeletons in formic acid, turning themselves into moving chemical hazards that burn anything trying to grab them.

The tarantula hawk wasp carries one of the strongest insect neurotoxins on Earth. Its sting temporarily paralyzes tarantulas while keeping their organs fresh for its larva to eat alive.

Tarantula Hawk Wasp
Pepsis spp.

COMMON LOCATION:
The Americas, especially warm desert and tropical regions.

DESCRIPTION:
Armed with one of the most powerful insect stings known, the tarantula hawk wasp delivers a neurotoxic sting that temporarily paralyzes tarantulas, keeping them alive and fresh as food for its developing larva.

Spitting spiders mix two chemicals during the shot itself.
They eject silk and venom simultaneously, which react mid-air to
create a sticky, fast-hardening glue net.

**Millipedes ooze cyanide-laced secretions when
threatened.** The amount isn't enough to kill a human but can
make predators foam, gag, or drop them instantly.

**Velvet ants (actually wasps) produce venom that affects
pain pathways directly.** Their sting is so intense it earned
them the nickname 'cow killers,' though the toxicity is mostly a
pain weapon, not a lethal one.

Some termites explode as a last resort. A special soldier
caste fills itself with sticky, toxic fluid that ruptures on attackers,
gluing and poisoning rival insects.

The giant hornet's sting dissolves tissue on contact. Its
venom contains chemicals that break down cellular structures,
causing severe pain and rapid spreading damage.

**Antlions paralyze prey with small amounts of neurotoxins
before dragging them under sand.** Their venom stops muscle
control, making escape impossible once the sand pit begins
collapsing.

**Some rove beetles weaponize a chemical that melts
through rival insects' armor.** Their defensive spray contains
alkaloids that break down waxy exoskeleton coatings, leaving
attackers exposed and dehydrated.

**The bombardier beetle's explosion is strong enough to
stun a frog.** The rapidly expanding gas inside its reaction
chamber creates a small shockwave that can physically knock
predators backward.

**Certain leaf beetles steal toxins from the plants they eat
and store them in their blood.** Predators that bite them get a
mouthful of the plant's own chemical defenses, repurposed like

stolen ammunition.

Trap-jaw ants spray formic acid into wounds they inflict.
The chemical intensifies the damage by preventing clotting and
amplifying irritation, making escape almost impossible for prey.

**The vinegarroon sprays pure acetic acid-strong enough to
burn skin.** It produces a mist of concentrated vinegar-like liquid
that stings the eyes and mucous membranes of predators.

Centipedes inject venom that dissolves muscle fibers.
Their venom contains metalloproteases, enzymes that break
down proteins and allow them to consume prey much larger than
themselves.

Centipede
Chilopoda

COMMON LOCATION:
Worldwide, especially in soil, leaf
litter, and under stones.

DESCRIPTION: Centipedes inject
venom containing enzymes called
metalloproteases that break down
muscle proteins, allowing them to
subdue and consume prey much
larger than themselves.

**Some caterpillars store cyanide from the plants they feed
on.** When attacked, they release the cyanide through tiny glands,
poisoning predators that try to swallow them.

**The puss caterpillar carries venomous spines that contain
thermoactivated toxins.** These compounds trigger intense
burning sensations that feel like a hot iron pressed against the
skin.

**Ants of the genus Crematogaster spray venom from their
stinger with surprising accuracy.** They can pivot their

abdomen over their head like a turret and fire venom directly at an enemy's face.

Certain spiders spit venomous glue in figure-eight patterns. The double-loop pattern allows them to hit moving prey with a sticky, paralyzing chemical net.

Net-casting spiders coat their elastic webs with a chemical that slows insect movement. Even fast fliers become sluggish the moment they touch the chemically treated strands.

The blue dragon sea slug steals stinging cells from jellyfish it eats. It stores them in its own tissues and uses them as a chemical defense far stronger than the jellyfish's original sting.

Some millipedes produce benzoquinones that can permanently stain skin. These chemicals react with sweat and oils, leaving dark marks that take weeks to fade.

Bullet ants produce one of the most painful venoms of any known insect. Their toxin, poneratoxin, interferes with nerve signaling, causing hours-long waves of intense burning.

Australian green ants spray a mist that contains antimicrobial chemicals. They use it not just as a defense but to disinfect their nests and even their larvae.

Some termites build chemical booby traps inside their tunnels. They line narrow passages with toxins so invading ants absorb chemicals while passing through.

Certain assassin bugs cover themselves in the corpses of ants they've killed. The decaying bodies release chemical signatures that mask the bug's real scent from predators and prey.

Tiger beetles spit digestive enzymes on prey before eating them. The enzymes soften the prey's exoskeleton, allowing the beetle to chew through armor-like structures easily.

The velvet worm sprays glue faster than a human blink. Its adhesive strands harden on contact with air, trapping insects instantly.

Some ants create formic acid "gas chambers" inside their nests. Workers flood chambers with acid vapor when threatened, suffocating intruders.

The emerald cockroach wasp injects venom directly into a roach's brain. The venom shuts down the roach's escape response, turning it into a chemically controlled zombie.

Certain diving beetles produce toxins that stun tadpoles and small fish. This allows them to hunt in both water and air, a rare ability in insects.

The bombardier beetle's chemicals reach 100°C (212°F). The explosion is so hot it creates audible popping sounds like miniature firecrackers.

Some tarantulas flick venom-coated hairs at predators. These urticating hairs cause chemical burns, itching, and inflammation on contact.

The Lycosa spider's venom contains compounds that numb pain. It paralyzes prey gently, keeping them alive and fresh longer.

Asian needle ants inject venom through a stinger shaped like a hypodermic needle. Their toxins spread rapidly, causing intense pain and localized tissue death.

Certain wasp larvae release chemicals that suppress a host's immune system. This keeps the host alive longer, allowing larvae to feed without interference.

Some ants can detect venom strength by smell. They judge rival colonies' chemical weapons long before a fight begins.

The bombardier beetle can fire its chemical jet multiple times in rapid bursts. Each shot requires only milliseconds to prepare inside its reaction chamber.

Bombardier Beetle
Brachininae

COMMON LOCATION:
Worldwide.

DESCRIPTION:
Bombardier beetles can fire repeated bursts of a boiling chemical spray, with each explosive shot prepared in milliseconds inside a specialized reaction chamber in their abdomen.

Giant water bugs inject digestive venom that liquefies muscle from the inside. This allows them to consume prey like fish, frogs, and even small snakes.

Some centipedes produce venom that blocks oxygen from entering cells. This chemical suffocation quickly immobilizes prey.

The Maricopa harvester ant has venom more toxic by weight than a cobra's. A single sting is painful, but dozens can overwhelm even large animals.

Certain millipedes create warning odors detectable meters away. These scents act as chemical alarms telling predators they taste terrible-or are poisonous.

Some flies regurgitate acidic digestive fluid onto their food. This dissolves the surface so they can sponge up nutrients through their mouthparts.

Ants in South America farm plants that produce toxins.
The ants use these toxic leaves as fortified walls to repel
herbivores.

**The zebra tarantula produces venom that causes sudden
muscle cramps.** The contractions immobilize prey almost
instantly.

**Some caterpillars puff out bright, toxic glands when
threatened.** These glands store defensive chemicals borrowed
from poisonous plants.

**The vinegarroon's spray can corrode metal when
concentrated.** Though harmless in small amounts, its acetic
acid secretion is chemically potent.

**Certain predatory beetles shoot foul chemicals from both
ends of their bodies.** This double-direction spray confuses
predators long enough for escape.

**Some scorpions glow under UV light due to chemicals in
their exoskeleton.** These compounds may help them detect
moonlight or avoid predators.

**Ants can create chemical footprints that mislead
predators.** They lay trails that lead away from the colony,
chemically tricking attackers.

**Some spiders dissolve prey externally before drinking
them.** Their venom contains enzymes that break down tissues
outside the body.

**Ichneumon wasps inject venom that freezes caterpillars
but keeps their hearts beating.** This allows larvae to feed on
living tissue without killing the host too soon.

**Certain millipedes curl into a ball and leak toxins onto
attackers.** The toxins irritate eyes, skin, and even animal
snouts.

MYTHS - BUSTED

People often believe that all brightly colored insects are poisonous. While many toxic species advertise danger with color, plenty of harmless insects mimic those patterns to fool predators. The chemistry, not the color, determines toxicity.

A common myth claims that venomous insects inject poison every time they bite or sting. Many species conserve their venom and use 'dry stings' or warning bites unless truly threatened, because venom is metabolically expensive to produce.

Some think that larger insects always have more powerful venom. Venom potency is unrelated to size-tiny ants like the Maricopa harvester ant possess toxins far more potent than many large spiders.

Many people assume millipedes can bite and inject venom. Millipedes cannot bite venomously; instead, they rely entirely on chemical secretions like cyanide compounds that deter predators through contact, not injection.

It's widely believed that scorpions glow under UV light because of their venom. The glow actually comes from chemicals in their exoskeleton, not their venom, and scientists still don't fully understand the purpose.

Some think bombardier beetles store their explosive chemicals premixed inside their bodies. The chemicals are stored separately and only become explosive when combined inside a specialized reaction chamber-preventing the beetle from blowing itself up.

LEGENDS

LEGENDS

In the rainforests of Borneo, Dusun hunters told of the 'Burning Beetle' whose breath could scorch skin. Early European naturalists wrote of encountering beetles that released a hot, stinging vapor when disturbed. Later studies confirmed bombardier beetles lived in the region, and their boiling chemical jets likely inspired generations of stories about insects that "breathed fire" when angered.

Along the rivers of northern Peru, villagers warned of the 'Death-Touch Spider,' a pale hunter said to stop hearts with a sting. Ethnographers recorded tales of livestock collapsing after spider encounters, though the truth was likely venom-induced paralysis from wandering spiders. The folklore persisted because the spider's pale color and silent movement made it appear ghostlike in lamplight.

In West Africa, elders spoke of the 'Night Perfume,' a millipede whose scent could make a grown man collapse. Travel diaries from the late 1800s mention caravans halting because of overwhelming odors from giant millipedes during the rainy season. The creatures released benzoquinones-chemicals that irritate eyes and lungs-leading to legends of invisible nighttime spirits that "steal the breath."

High in Japan's Kiso Mountains, villagers whispered about the 'Mist Scorpion' whose glow foretold illness. Researchers later discovered scorpions fluoresce under moonlit conditions when dew is present. Shepherds who fell sick during humid seasons associated the glowing shapes with disease, though the creatures were harmless unless stepped on.

In the Australian outback, old cattlemen spoke of the 'Sting of Silence,' an ant that left no wound but caused hours of agony. Modern biology solved the mystery: bullet ants and their relatives deliver venom that disrupts nerve signals without leaving visible marks. Early ranch hands thought the land itself was cursed, since they never saw what stung them.

In the Amazon Basin, river guides warned of the 'Sleeping Roach,' a creature said to hypnotize prey with a glance. This legend traces back to the emerald cockroach wasp, which injects venom directly into the brain of roaches, shutting down their escape reflex. Locals observing zombie-like roaches believed the wasp cast a spell rather than delivering a precise neurological strike.

In the cloud forests of Costa Rica, stories persist of the 'Silver Thread Spirit' that binds animals in sticky ropes at night. Field researchers later linked the legend to velvet worms, which spray fast-hardening adhesive that glitters under moonlight. Animals found tangled in shimmering strands fueled tales of a forest spirit weaving silver traps.

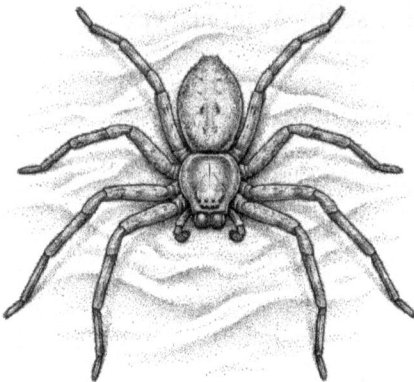

Death Touch Spider
Sicarius spp.

COMMON LOCATION: Deserts of southern Africa and South America.

DESCRIPTION: This sand-dwelling spider delivers venom that damages tissue and blood cells, earning its nickname for the powerful effects of a single bite, while spending most of its life hidden beneath the sand.

Poison Dart Caterpillar
Lepidoptera (toxin-sequestering larva)

COMMON LOCATION: Tropical regions.
Stores plant toxins in its body to deter predators.

DID YOU KNOW?

Did you know? Some ants can detect the chemical signature of venom before a fight even begins. Their antennae read the volatile compounds drifting off rival colonies, letting them decide whether to battle, retreat, or call for reinforcements long before any physical contact happens.

Did you know? Certain beetles are born with detoxifying enzymes that neutralize plant poisons their parents ate. This inherited chemical shield allows them to feed on toxic plants immediately after hatching, giving them access to food sources other insects avoid entirely.

Did you know? Some millipedes produce cyanide compounds strong enough to kill small predators but mild enough that birds sometimes roll in them intentionally. Researchers think the birds may be using the chemicals as parasite control, effectively turning the millipede's poison into a natural insect repellent.

Did you know? Scorpions adjust the potency of their venom depending on their target. They conserve their most expensive toxins for subduing tough prey, while defensive stings often use a diluted mix-saving their strongest chemistry for when it really counts.

Did you know? Many spiders can change the chemical composition of their venom across their lifetime. Younger spiders rely more on fast-acting neurotoxins to secure small prey, while adults shift to enzymes that break down larger animals more efficiently.

🐞 **Did you know?** Some caterpillars store defensive toxins in specific sacs that puff outward only when threatened. This keeps their chemistry hidden until the exact moment it's needed, surprising predators with sudden bursts of color and poison.

🐞 **Did you know?** Certain aquatic insects inject venom that disrupts oxygen flow in tissues. This allows them to overpower frogs, tadpoles, and even small fish, using chemistry instead of strength to dominate underwater battles.

🐞 **Did you know?** A few species of ants mix venom with soil minerals to create a more abrasive sting. This gritty chemical blend tears into soft tissue more effectively, amplifying the burn and making the wound harder to ignore.

🐞 **Did you know?** Some spiders inject a venom that begins digesting prey before the bite is even withdrawn. The enzymes spread rapidly under the skin, turning muscle fibers into liquid that the spider later drinks like broth.

Velvet Ant
Mutillidae family

COMMON LOCATION: Worldwide.
Delivers one of the most painful insect stings despite its harmless appearance.

STORY MOMENT

STORY MOMENT

The Night the Forest Breathed Fire

Dr. Lena Moretti had studied bombardier beetles for years, but nothing prepared her for hearing one launch its chemical blast echo across an open rainforest valley. The sound arrived first-a sharp pop like a snapped twig-followed by a brief shimmer of heat that rippled in the humid air. When she swept her headlamp toward the noise, she caught a glimpse of the beetle standing its ground against a night lizard nearly ten times its size. A second pop burst from its abdomen, this time with a tiny cloud of steam twisting upward like smoke from a miniature engine.

The lizard recoiled, shaking its head violently as droplets of the beetle's boiling benzoquinone mixture peppered its snout. Lena watched in awe: the beetle's reaction chamber worked faster than any chemical device she had ever seen, mixing two unstable compounds only milliseconds before detonation. Each blast pushed the predator farther back until it finally turned and vanished into the undergrowth, leaving the beetle victorious in a battle it should have had no chance of winning.

Later, as she knelt beside the beetle, recording its posture and the faint crackle still coming from its abdomen, Lena felt a deeper truth settle in. Nature wasn't just creative-it was engineered with a precision no laboratory could replicate. Insects like this weren't accidents of evolution-they were proof that some of the most extraordinary chemical weapons on Earth were hidden in plain sight, carried by tiny creatures most people never noticed at all.

MIND-BLOWN™ Cartoons

Bug Smiles

Q. What did one flea say to the other flea?
A. Should we walk or take a dog?

Q. What do you call a fly with no wings?
A. A walk.

Q. Why was the centipede late for school?
A. He had to tie his shoes.

Q. What did the caterpillar say to the butterfly?
A. You've changed.

FUN QUIZ

1. Which insect fires a boiling chemical blast as its primary defense? A) Fire ant   B) Bombardier beetle   C) Assassin bug   D) Velvet ant

2. What does the emerald cockroach wasp's venom shut down in its prey? A) Heart rate   B) Escape reflex   C) Vision   D) Digestion

3. Which creature sprays fast-hardening glue to trap prey?   A) Spitting spider   B) Tiger beetle   C) Trap-jaw ant   D) Velvet worm

4. True or False: Millipedes inject venom through biting.

5. True or False: Bullet ant venom causes one of the most painful stings in the insect world.

6. Which insect uses formic acid as both a weapon and a disinfectant? A) Australian green ant   B) Firefly larva   C) Net-casting spider   D) Pseudoscorpion

7. What do assassin bugs inject to liquefy their prey?   A) Sleeping toxins   B) Neurotoxins   C) Digestive enzymes   D) Alkaloids

8. True or False: Scorpions glow because their venom contains fluorescent chemicals.

9. Which predator dissolves prey from the outside before consuming it? A) Antlion   B) Wolf spider   C) Spitting spider D) Many spiders

10. What chemical do some millipedes release that can stain skin for weeks? A) Formic acid   B) Benzoquinones   C) Hydroquinone   D) Cyanide gas

QUIZ ANSWERS

1: B) Bombardier beetle

2: B) Escape reflex

3: D) Velvet worm

4: False

5: True

6: A) Australian green ant

7: C) Digestive enzymes

8: False

9: D) Many spiders

10: B) Benzoquinones

CHAPTER 3

3- Zombie Makers & Mind-Control Monsters

Parasites, fungi, behavioral hijacking

Emerald Cockroach Wasp
Ampulex compressa

COMMON LOCATION: Africa, Southeast Asia
Stings cockroaches to control their movement and uses them as living hosts for its larvae.

FUN & WEIRD FACTS

A zombie fungus called Ophiocordyceps can take control of an ant's legs and jaws. It releases chemicals that force the ant to climb a plant and clamp down tightly. This puts the ant in the perfect spot for the fungus to grow and release spores.

The jewel wasp turns a cockroach into a willing "pet" by injecting venom into its brain. The venom shuts down the roach's escape instinct but keeps it alive. This lets the wasp walk the roach around by its antenna like a leash.

Hairworms hijack crickets and make them jump into water-even though crickets drown there. The worm releases chemicals that override the cricket's fear of water. The worm then escapes into the water to continue its life cycle.

The lancet liver fluke makes ants climb grass at night and clamp down tightly. Cold temperatures activate the parasite's control chemicals. During the day the ant behaves normally, preventing overheating of the parasite.

Parasitoid phorid flies lay eggs that eventually decapitate ants-but only after changing their behavior first. The larva makes the ant wander away from the colony before dying. This prevents other ants from detecting and removing the infected host.

A parasite called Leucochloridium turns snail eye stalks into bright, pulsing "worms." The pulsation attracts birds, the parasite's next host. The snail effectively becomes a glowing billboard for infection.

Some fungi infect insects and remove their fear, making them walk into sunlight or open spaces. The fungus manipulates neurotransmitters linked to caution and hiding. This exposes the insect so spores spread farther.

Tachinid fly larvae can steer caterpillars while feeding on them. They release hormones that alter the caterpillar's defensive behavior. Infected caterpillars sometimes protect the parasite instead of themselves.

Gall wasps trick plants into growing strange bulb-like structures for their young. The chemical cocktail they inject rewrites plant growth patterns. These "galls" act like custom-built houses for the larvae.

GALL WASP
Family Cynipidae (adult)

COMMON LOCATION:
Woodlands, meadows, and shrublands worldwide, especially on oak, rose, and willow plants.

DESCRIPTION:
Gall wasps survive by rewriting plants from the inside out. When a female lays her eggs, she injects a precise chemical cocktail that hijacks the plant's growth system, forcing it to form a swollen, bulb-structure called a gall. The gall becomes a custom-built nursery—providing food, shelter, and protection–while the developing larva safely grows inside living plant tissue shaped entirely for its needs.

A microsporidian parasite makes ants cluster together unnaturally. The parasite alters social scent signals that ants rely on. Crowding helps the parasite spread rapidly through the colony.

Certain mites latch onto spiders and stop them from building webs. They interfere with the spider's motor patterns and timing. The spider's halted behavior helps the mite feed safely.

The parasitoid wasp Euderus set forces its host insect to block its own escape hole. The wasp larva alters the host's digging behavior, trapping it in place. This gives the wasp a sealed, protected chamber to mature inside.

A fungus infecting cicadas pumps them with stimulant-like chemicals. The infected cicadas walk and fly frantically, spreading spores farther. To the cicada, the chemicals feel like an unstoppable urge to move.

Nematomorph worms make grasshoppers leap toward predators. They disturb the host's balance and decision-making centers. The worm needs the host to be eaten to complete its life cycle.

Certain bacteria infect aphids and erase their instinct to flee. The infection alters their neural escape circuits. Predators asily pick them off, helping spread bacteria to ecosystem.

A wasp larva can force spiders to rebuild their webs in a totally new pattern. The altered web design protects the wasp's cocoon. Scientists call these structures "zombie webs."

A fly-killing fungus forces its hosts to die at sunset. Humidity and temperature at dusk are ideal for spore release. The fungus times death with surprising precision.

A tiny worm parasite turns black ants bright yellow or orange. Predators mistake them for berries and eat them. The parasite then continues its life cycle in the predator.

A beetle larva can chemically hypnotize a bee into guarding it. The bee acts like a bodyguard, even as the larva feeds. This behavior serves the parasite at the cost of the bee's survival.

Some trematode parasites alter fish movement so they swim near sunlight. This increases the chance of predators spotting them. The parasite needs the predator to complete its life cycle.

A "zombie cicada" fungus dissolves the insect's abdomen but keeps it walking. It releases compounds similar to amphetamines that overstimulate the nervous system. Infected cicadas walk around shedding spores like salt from a shaker.

Caterpillars infected by certain wasps thrash violently at anything nearby. This protects the wasp's cocoon hanging below the caterpillar. The caterpillar becomes a mind-controlled guard dog.

A barnacle parasite transforms crabs into full-time babysitters. It alters their hormones so they behave like brooding mothers-even males. The crab cares for barnacle offspring as if they were its own.

Some brain parasites change the temperatures ants prefer. The ants begin seeking warm or cool spots perfect for parasite development. To observers, the ants seem strangely "lost."

Certain viruses infect caterpillars and make them climb tall plants before dying. After death, the caterpillar liquefies, releasing millions of virus particles downward. The virus essentially uses gravity as a delivery system.

A wasp larva forces a ladybug to become a frozen bodyguard. Neurotoxins make the ladybug stand still for weeks. The larva pupates beneath it, protected by the living "shield."

Some fungi infect spiders and force them to build webs they would never make. These webs support fungal growth instead of catching prey. The spider dies hanging beneath the structure like an ornament.

Certain nematodes alter insect scents to mimic smells predators prefer. This increases the insect's chance of being eaten. The parasite uses the predator as a new host environment.

Tiny parasitic isopods hijack shrimp behavior and make them swim near the surface. This heightens their chance of being eaten by birds. The parasite then uses the bird to complete its life cycle.

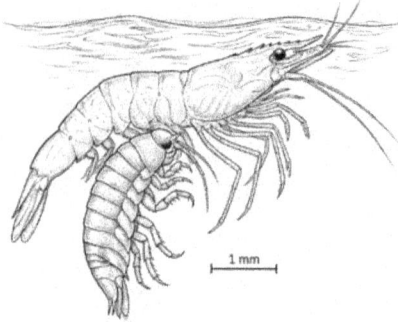

PARASITIC ISOPOD
Ichthyoxenus sp.

Tiny parasitic isopods hijack shrimp behavior and make them swim near the surface. This heightens their chance of being eaten by birds. The parasite then uses the bird to complete its life cycle.

1 mm

Mind-control fungi can rewrite insect day-night cycles. Infected insects become active at odd times that benefit the fungus. This helps spores survive and spread.

Wasp larvae interfere with a caterpillar's immune system. This keeps the host alive but unable to heal. The parasite gets a predictable, long-term food source.

Worm parasites inside crickets cause exaggerated jump reflexes. They stimulate neural pathways linked to sudden movement. This chaotic jumping behavior draws predators.

A tropical fungus changes beetles' sense of smell. The beetles start seeking odors linked to fungal growth, not food. This creates a self-spreading infection loop.

Some mites damage bees' chemical sensors so bees wander off-course. Disoriented bees spread mites across greater distances. Colonies weaken as more workers become infected.

A wasp venom can erase a spider's short-term memory. The spider forgets its normal web patterns. It rebuilds a simpler web ideal for the wasp's cocoon.

Nematodes infecting ants make them stand still with raised abdomens. Workers mistake them for queens and feed them. This spreads the parasite through oral contact.

A protozoan parasite makes beetles climb plants before bursting out. The height favors wind dispersal. The beetle becomes an elevator for parasite release.

Certain fungi cause flies to spread their wings in a dramatic "death pose." This opens more surface area for spore release. The pose looks ritualistic but is pure biology.

Nematode-infected flies seek windy places. High winds carry the nematodes to new environments. The behavior is completely unnatural for a healthy fly.

Some parasitoid wasps inject venom that stops feeding behavior entirely. The host becomes weak and slow-but alive. This ensures the wasp larva has no competition.

Worm parasites make ants stand motionless for long periods. Stillness protects the parasite during sensitive development stages. Healthy ants rarely freeze like this.

A fungus infecting butterflies disrupts wing coordination. The butterfly flies in zigzags and spirals. These odd patterns spread spores mid-flight.

Parasitoid fly larvae make ants walk in looping patterns. The odd movements confuse other ants. Loops help spread the parasite without triggering alarm pheromones.

Viral infections can freeze caterpillars in oversized, swollen forms. Their hormones stop progressing to the next stage. This gives the virus more tissue to use for replication.

Barnacle parasites can dial down crab aggression. The crab stops defending territory. This makes caring for the parasite easier.

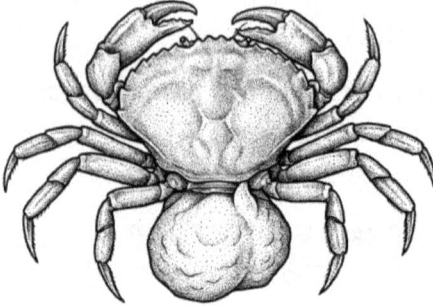

Parasitic Barnacle
Sacculina spp.

COMMON LOCATION:
Coastal waters worldwide.

DESCRIPTION:
Parasitic barnacles can alter a crab's hormones and behavior, reducing aggression and territorial defense so the crab instead protects and cares for the parasite as if it were its own eggs.

Nematodes drive caterpillars upward, then kill them at height. Sun exposure dries the corpse, preserving nematodes until new hosts arrive. It's nature's version of a biological billboard.

A wasp's brain-targeting venom disables cockroach movement but leaves thinking intact. The roach knows it's being controlled but cannot resist. Scientists compare it to chemical handcuffs.

Fungi can alter how ants respond to gravity. Infected ants climb downward or sideways against instinct. This puts them in microclimates the fungus prefers.

Some protozoa hijack insect mating signals. The host acts unusually flirty, attracting more insects to infect. It turns reproduction into a transmission tool.

Mites infecting dragonflies slow their wingbeats. This makes dragonflies easier prey. The predator becomes the mite's next ride.

MIND CONTROLLING INSECTS

1 HAIRWORM
Spinochordodes tellinii

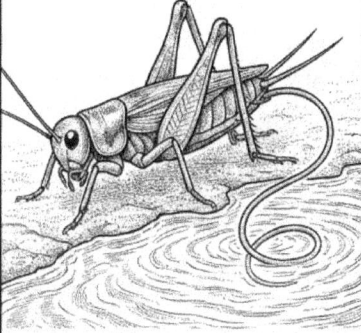

COMMON LOCATION: Fresh water
Short, vivid description: Cricket seeks water, manipulated by the parasite to release itself into the aquatic environment.

2 PHORID FLY
Pseudacteon spp.

COMMON LOCATION: Tropical forests
Short, vivid description: Fly injects eggs, causing ant to lose coordination and wander aimlessly before decapitation.

3 ZOMBIE ANT FUNGUS
Ophiocordyceps unilateralis

COMMON LOCATION: Rainforests
Short, vivid description: Fungus forces ant to climb and bite onto vegetation, securing a position for spore dispersal.

4 JEWEL WASP
Ampulex compressa

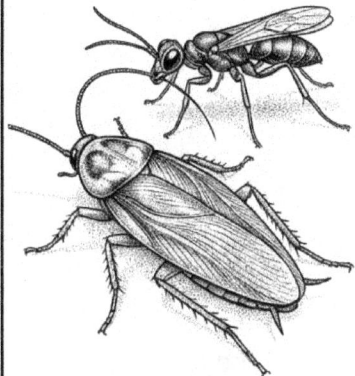

COMMON LOCATION: Tropical regions
Short, vivid description: Wasp guides pacified cockroach to its burrow to serve as living food for its larva.

MYTHS - BUSTED

Many people think zombie fungi instantly control their hosts like puppets. In reality, the fungus grows slowly through the insect's body and alters behavior over days. The "instant takeover" seen in videos is just the final stage.

Some believe parasites always kill their hosts quickly. Most mind-controlling parasites keep the host alive as long as possible. A living host spreads infection better than a dead one.

It's a common myth that all parasitic wasps paralyze their hosts permanently. Some wasp venoms selectively disable only certain behaviors, like escape reflexes. The host stays alive and mobile to benefit the wasp's developing larva.

People often assume mind-control parasites only target insects. Fungi, worms, protozoa, viruses, and mites all use behavior manipulation. The strategy evolved independently many times across tiny creatures.

A lot of readers think mind-control always involves brain infection. Many parasites never touch the brain at all-they alter hormones, muscles, or chemical signals. Hosts act like zombies even without direct brain damage.

Some believe infected insects know they're being controlled. Most behavior changes happen because natural instincts are chemically altered. The insect simply "feels" compelled, not aware of manipulation.

There's a myth that zombie fungi work the same on every insect. Different species respond differently, and many insects can't be infected at all. Mind control in nature is highly specialized.

LEGENDS

LEGENDS

In the mountains of Vietnam, villagers told of the "Night Walker Ants" that climbed trees in perfect unison before dying. Travel journals from the early 1900s described long lines of ants ascending trunks at dusk, all biting the undersides of leaves before freezing in place. Locals believed a forest spirit summoned them, but scientists later found the same pattern seen today in Ophiocordyceps outbreaks-behavior-altering fungi quietly directing their final steps.

Amazon hunters spoke of the "Restless Cicadas," insects said to wander the forest long after their bodies began to rot. Ethnographers recorded descriptions that mirror the modern zombie cicada fungus, which destroys the abdomen yet floods the insect with chemicals that keep it moving. The tribes believed the cicadas were carrying forest messages to unseen spirits, never realizing a fungus was pulling the strings.

Along the coast of Japan, fishermen whispered about crabs that acted like mothers even when they were male. Accounts from the Edo period describe crabs guarding egg-like masses and refusing food for weeks. These behaviors match today's known effects of parasitic barnacles, which hijack crab hormones and turn any host-male or female-into a nurturing caretaker for the parasite's brood.

In the rainforests of Peru, elders spoke of the "Lost Guardians," caterpillars that violently protected empty spaces in the night. Stories tell of caterpillars thrashing at shadows as if defending something unseen. Explorers later discovered parasitic wasps in the region whose larvae implant neurochemicals that turn the host into a bodyguard for the developing cocoon-a defense so fierce it inspired tales of enchanted warriors.

In rural Kenya, older generations warned of the "Wandering Termites," workers that left their colonies and walked far into the grasslands before collapsing. Records from early entomologists described lone termites found miles from any mound, as if compelled by an invisible force. Modern science recognizes this as classic fungal manipulation: infected termites are pushed away from the colony to prevent detection while spreading spores across new terrain.

High in Costa Rica's cloud forests, guides spoke of beetles that climbed trees only to fall like rain. Travelers described beetles scaling trunks again and again, even after falling repeatedly. These legends parallel infections by parasitic protozoa and fungi that alter climbing behavior, driving beetles upward to improve spore dispersal. The bizarre repetition made locals believe the beetles were obeying a forest curse.

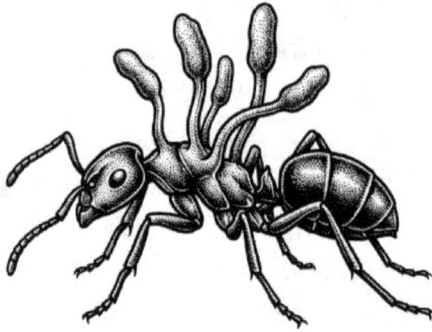

Cordyceps-Infected Ant
Ophiocordyceps spp.

COMMON LOCATION: Tropical forests worldwide
Fungal parasites alter ant behavior to spread spores and complete their life cycle.

Bug Smiles

Q. Why did the ant stop making its own decisions?

A. It outsourced them to a fungus with *strong leadership skills*.

Q. Why don't zombie-making parasites ever need a remote control?

A. Because their hosts are already on *manual override*.

Q. Why did the caterpillar feel unusually motivated one morning?

A. It realized the debate was *already decided*.

Q. Why do mind-control insects never panic?

A. Someone else is always *thinking for them*.

DID YOU KNOW?

🪲 **Did you know?** Some fungi can release chemicals that mimic an insect's own neurotransmitters, tricking the host into believing the parasite's commands are its own instincts. This chemical disguise is so convincing that infected insects often continue performing normal tasks even while their bodies are being overtaken. Scientists compare it to getting hacked without realizing anything is wrong.

🪲 **Did you know?** Jewel wasps target the exact neural cluster in a cockroach that controls "escape decisions," not movement. This means the roach can walk, climb, and move normally-yet it feels no motivation to flee danger. The parasite doesn't need to paralyze the host; it only needs to silence the will to resist.

🪲 **Did you know?** Hairworms use light and reflection cues to lure insects toward water. Their chemicals make the host interpret shimmering surfaces as safe pathways instead of deadly traps. This altered perception is so strong that insects will drown themselves in puddles they normally avoid.

🪲 **Did you know?** The famous "zombie ant grip" is created by a fungus hijacking the ant's jaw muscles directly-not its brain. The fungus replaces muscle tissue with fungal fibers that force the jaw shut at the exact moment the parasite needs. Even after the ant dies, the grip stays locked like a natural clamp.

Did you know? Some viruses rewrite a caterpillar's molting schedule, trapping it in a super-sized stage perfect for viral growth. The altered hormones make the caterpillar eat nonstop, producing more tissue for the virus to infect. When it finally dies, the extra mass becomes a massive viral release chamber.

Did you know? Parasitic barnacles inject chemicals that override crab hormones, convincing the crab that it is carrying eggs-even if it's a male. The crab then fans, protects, and cleans the parasite's brood as if they were its own young. This is one of the few cases where a parasite alters parental behavior across sexes.

Did you know? Some parasitic flies release chemicals that distort an ant's internal clock. The infected ant becomes active at strange hours, wandering away from the colony while other ants are inactive. This solitary behavior helps protect the parasite developing inside the ant's body.

Did you know? Certain worms can manipulate insect posture by chemically locking specific muscles. Ants infected by these parasites raise their abdomens unnaturally, mimicking queen ants and attracting special care from workers. This spreads the infection through feeding and grooming behaviors.

Did you know? Fungi use environmental cues-like humidity, temperature, and sunlight angle-to time exactly when they kill their host. This ensures spores launch when wind and moisture conditions are perfect for spreading. It's so precise that infected insects often die within the same one-hour window every day.

ZOMBIE MAKERS & MIND-CONTROL MONSTERS
Real creatures that hijack insect behavior

Hairworm-Controlled Cricket
Nematomorpha spp.

COMMON LOCATION: Worldwide.
A parasitic worm alters the cricket's behavior,
forcing it to seek water.

Phorid Fly Ant
Phoridae family

COMMON LOCATION: Worldwide.
Parasitic flies alter ant behavior while
developing inside the host.

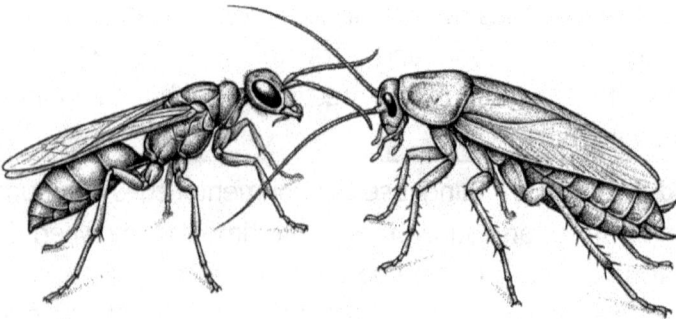

Emerald Cockroach Wasp
Ampulex compressa

COMMON LOCATION: Africa and parts of Asia.
Injects venom into a cockroach's brain, turning it into a living food supply for its young.

STORY MOMENT

STORY MOMENT

The Night of the Silent March

Dr. Aiden Wells had followed army ants through rainforests for years, but he had never seen them like this-moving in a perfectly straight line at midnight, silent, slow, and utterly wrong. Army ants normally flowed like a living river, bursting with motion and aggression. But these ants marched as if being guided by something unseen, each one lifting and lowering its legs in the same eerie rhythm.

He knelt closer, sweeping his red-light lamp across their bodies, and the truth hit him like a punch. Tiny white filaments clung to the joints of the ants-the early threads of a parasitic fungus beginning to bloom. The ants weren't following their queen or responding to pheromone trails. They were following fungal chemicals that had hijacked their instincts, turning the colony's strongest workers into perfect delivery vehicles. The ants weren't marching home-they were marching to die.

Aiden watched as the first ant broke formation and began climbing a low branch, moving slowly but with unwavering purpose. Others followed, one by one, ascending the trunk like sleepwalkers. In the dark, the forest felt impossibly still. He knew that by sunrise, these ants would be frozen in place, jaws locked, bodies still, as the fungus erupted from their exoskeletons. It was breathtaking and terrifying all at once-the realization that some of the world's most astonishing mind-control strategies weren't science fiction at all, but living, growing, unfolding right in front of him.

ZOMBIE MAKERS

GREEN-BANDED BROODSAC
Leucochloridium paradoxum

COMMON LOCATION: Temperate forests and gardens.
Short, vivid description: Parasite invades snail's eye stalks, making them pulsate to attract birds, forcing the host into open areas.

PARASITOID WASP LARVAE
Cotesia glomerata

COMMON LOCATION: Gardens and agricultural fields.
Short, vivid description: Wasp larvae emerge from living caterpillar, which then spins a protective web and guards the cocoons until it dies.

ZOMBIE ANT FUNGUS
Ophiocordyceps unilateralis

COMMON LOCATION: Tropical rainforests.
Short, vivid description: Fungus manipulates ant to climb vegetation and bite into a leaf before killing it, positioning the body for optimal spore dispersal.

VARROA MITE
Varroa destructor

COMMON LOCATION: Beehives worldwide.
Short, vivid description: Mites attach to adult bees and larvae, feeding on fat bodies and transmitting viruses, weakening the host and colony.

FUN QUIZ

1. Which parasite forces ants to climb plants and lock their jaws in place?

> A) Gall wasp B) Ophiocordyceps fungus C) Phorid fly
> D) Jewel wasp

2. What behavior does the jewel wasp shut down in cockroaches?

> A) Vision B) Movement C) Escape reflex D) Breathing rhythm

3. Which parasite drives insects to jump into water?

> A) Hairworm B) Barnacle parasite C) Viral caterpillar infection D) Trematode worm

4. **True or False:** Some fungi time their host's death to occur at sunset?

5. **True or False:** Barnacle parasites can alter a crab's hormones so it behaves like a mother, even if it's male?

6. Which parasite causes snail eye stalks to pulse like worms to attract predators?

> A) Microsporidian B) Leucochloridium fluke C) Tachinid larva D) Mite

7. Which parasite can erase a spider's short-term memory and alter its web pattern?

> A) Jewel wasp B) Zombie fungus C) Parasitic wasp
> D) Protozoan

QUIZ ANSWERS

1: B) Ophiocordyceps fungus

2: C) Escape reflex

3: A) Hairworm

4: True

5: True

6: B) Leucochloridium fluke

7: C) Parasitic wasp

CHAPTER 4

4-Superpowers of the Bug World

Physical abilities: speed, vision, hearing, flight, strength

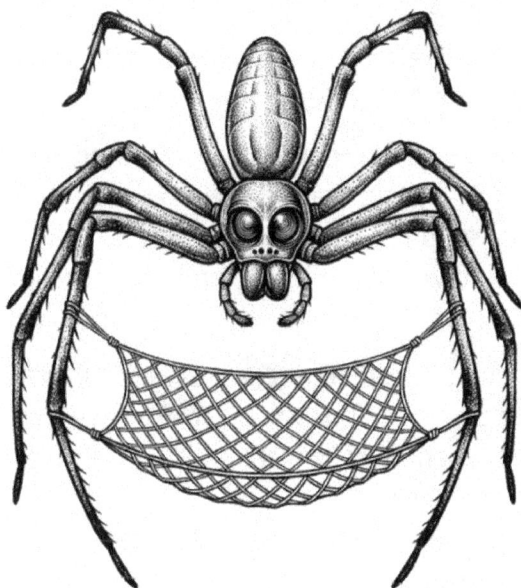

Net-Casting Spider (Ogre-Faced Spider)
Deinopis spp.

COMMON LOCATION: Tropical regions worldwide
Uses handheld silk nets and night vision to ambush prey.

MIND-BLOWN MOMENT

FUN & WEIRD FACTS

Dragonflies can predict the future position of their prey. They calculate interception points like a fighter jet, adjusting their flight path in milliseconds, which gives them an astonishing hunting success rate of up to 95%.

Trap-jaw ants snap their mandibles at over 140 miles per hour. They use this explosive strike to catch prey or launch themselves into the air like spring-loaded popcorn, and these tropical ants perform one of the fastest movements ever recorded in the insect world.

Click beetles (common beetles found across North America and Europe) launch themselves into the air without using their legs. They snap a hinge-like spine inside their body that blasts them upward with a loud click, and they can flip several inches high-like a human doing a 60-foot backflip from a lying position.

Some tiny shrimp-like creatures in shallow tropical seas strike fast enough to boil water. Their claws create shock waves called cavitation as they slam shut with incredible speed, and studying this power helps scientists understand similar rapid-strike mechanics in certain beetles and ants.

Tiger beetles (ultra-fast hunters found in sandy habitats worldwide) run so fast they temporarily go blind. Their eyes can't gather enough light at top speed, forcing them to pause to "reload" their vision, yet they are still among the fastest insects alive.

The ironclad beetle (a nearly uncrushable ground beetle from the U.S. Southwest) can survive being run over by a car. Its shell has interlocking layers that flex instead of crack under pressure, inspiring engineers who study its armor for designing more resilient protective gear.

Springtails (tiny soil jumpers found almost everywhere on Earth) can leap over 100 times their body length. They snap a tail-like structure called a furcula that fires them into the air like a biological catapult, making them some of the greatest pound-for-pound jumpers in the animal kingdom.

Water boatmen (tiny freshwater insects found in ponds, lakes, and streams worldwide) produce one of the loudest sounds on Earth for their size. They create their mating call by rubbing their genitals against their abdomen in a process called stridulation, and if they were human-sized their underwater "song" would be louder than a stadium concert.

Leafcutter ants can carry 20–50 times their own body weight. If humans matched this strength, we could lift a small car over our heads, and their specialized muscles let them haul enormous leaf pieces back to their underground fungus farms.

Honeybees can see ultraviolet patterns hidden from human eyes. Many flowers have UV "landing strips" that guide bees straight to nectar like glowing runway lights, giving honeybees an incredible foraging advantage.

Honeybee
Apis mellifera

COMMON LOCATION:
Worldwide

DESCRIPTION:
Honeybees can see ultraviolet patterns that are invisible to human eyes. Many flowers have hidden UV "landing strips" that guide bees straight to nectar like glowing runway lights, helping them find food quickly and efficiently.

Fire-chaser beetles (heat-sensing beetles found near burn zones) can detect forest fires from over 50 miles away. They use specialized infrared sensors that allow them to "see" heat, and they race toward burned areas because freshly charred trees are perfect places to lay eggs.

Praying mantises have true 3D depth perception. They combine signals from both eyes to judge distance with human-like accuracy, and this rare insect superpower gives them exceptional precision when ambushing prey.

Praying Mantis
Mantodea

COMMON LOCATION: Worldwide, especially tropical and temperate regions.

DESCRIPTION: Praying mantises have true **3D depth perception,** combining signals from both eyes to judge distance with remarkable accuracy. **This rare insect superpower allows them to strike with exceptional precision** when ambushing prey.

Bark beetles (tiny wood-boring beetles found in forests worldwide) can chew through solid wood despite being smaller than a grain of rice. Their jaws contain hardened metals like zinc that act like mini power tools, allowing them to tunnel fast enough to reshape entire forest ecosystems.

Army ants build living bridges using their own bodies. They link legs and jaws together to create structures that adjust in real time as the colony moves, letting the swarm behave like a single shape-shifting organism.

Thorn bugs (treehoppers with dramatic spines found in tropical forests) mimic thorns with near-perfect accuracy. Their camouflage is so convincing that predators ignore them even when they're in plain sight, giving them an armor-like disguise that keeps them safe.

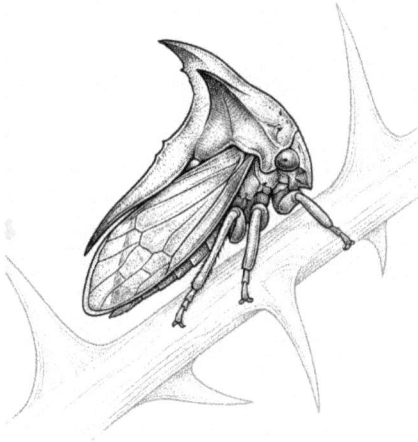

Thorn Bug
Membracidae family (adult)

COMMON LOCATION: Tropical and subtropical forests, especially Central and South America.

DESCRIPTION: The thorn bug looks less like an insect and more like a weaponized plant spine. Its hardened, sharply pointed body perfectly mimics the jagged thorns of the stems it lives on, making predators hesitate—or pass it by entirely. Birds scanning for soft prey often ignore thorn bugs completely, mistaking them for nothing more than a dangerous piece of the plant itself.

Bombardier beetles (found on every continent except Antarctica) can fire boiling chemical blasts from their rear end. They mix two chemicals in a reaction chamber that instantly heats the spray, giving them a rapid-fire defense system.

Dragonflies can reverse direction midair faster than a hummingbird. Their four wings move independently, allowing extreme agility that makes them elite aerial hunters.

Assassin bugs (fast-stabbing predators found in warm regions worldwide) have mouthparts that work like a biological syringe. They pierce prey, inject digestive enzymes, and drink the liquefied insides.

The Hercules beetle (a rhinoceros beetle from Central and South America) can lift over 850 times its own body weight. That's like a human lifting six SUVs at once.

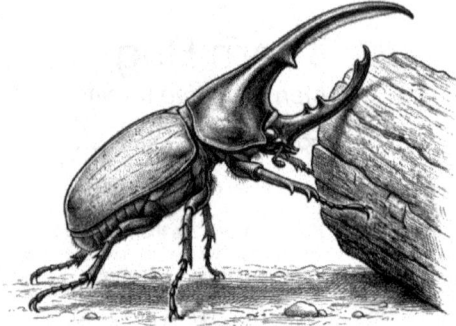

Hercules Beetle
Dynastes hercules

COMMON LOCATION:
Central and South America.

DESCRIPTION:
The Hercules beetle can lift and carry more than 850 times its own body weight. Scaled to humans, that would be like lifting six SUVs at once—making it one of the strongest animals on Earth for its size.

Mosquitoes can detect carbon dioxide from over 100 feet away. They track breath chemicals, then switch to heat and odor cues to lock onto humans with scary accuracy.

Houseflies process visual information about seven times faster than humans. This slow-motion perception helps them dodge swats with ease.

Housefly
Musca domestica

COMMON LOCATION:
Worldwide.

DESCRIPTION:
Houseflies process visual information about seven times faster than humans. This rapid vision creates a slow-motion view of the world, allowing them to react quickly and dodge swats with ease.

The atlas moth (a giant silk moth from Southeast Asia) has wing tips that resemble snake heads. This mimicry scares predators and buys the moth time to escape.

Antlion larvae (pit-digging predators found in sandy habitats worldwide) build collapsing sand traps. When insects fall in, the walls crumble, and escape becomes almost impossible.

Fireflies create light with nearly 100% efficiency. Their bioluminescence produces almost no heat, making them more efficient than human-made bulbs.

Spittlebugs (small plant-feeding insects found in grassy areas) can launch themselves with more power than any known animal relative to size. Their hydraulic legs fire them upward with explosive force.

Honeybees can sense Earth's magnetic field. This built-in compass guides them on long foraging trips.

The great diving beetle (a freshwater beetle found across Europe and Asia) carries air underwater like a scuba tank. It traps a bubble beneath its wings and refreshes it as it moves.

Ants can farm aphids like tiny cattle. They protect them in exchange for sweet honeydew.

Robber flies (aggressive aerial hunters found worldwide) catch insects midair with extreme precision. Their fast acceleration and sharp mouthparts make them the raptors of the insect world.

Some butterflies taste with their feet. Sensors on their legs help them choose the right plants for laying eggs.

The rainbow stag beetle (a brilliantly colored beetle from Australia) creates metallic colors using microscopic shell layers. The colors shift as light hits different angles.

Ants form floating rafts during floods. They interlock bodies to create waterproof platforms that last for weeks.

Raft-Building Ants
Solenopsis spp. (fire ants)

COMMON LOCATION: Flood-prone regions worldwide.

DESCRIPTION: When floods strike, some ants link their bodies together to form floating rafts that can survive for weeks. By interlocking legs and trapping air, the colony creates a living, waterproof platform that protects workers, queens, and larvae until land is reached.

Bug Smiles

I asked an ant why it was running in circles instead of carrying food. The ant said, "I'm not running in circles. I'm following instructions."

I asked who gave the instructions.
The ant said, "An ant I've never met, who got them from another ant, who probably misunderstood them."

Then it paused and added, "But if I stop, the ants behind me will panic, the ants ahead of me will keep going anyway, and somehow this will still be blamed on me."

It shrugged and said, "Welcome to the colony."

BUGS WITH TRUE SUPERPOWERS

Rhinoceros Beetle
Dynastinae

COMMON LOCATION: Tropical and subtropical regions worldwide. Can lift and push loads over 800 times its own body weight, making it one of the strongest animal on Earth relative to size.

Ironclad Beetle
Phloeodes diabolicus

COMMON LOCATION: WESTERN NORTH AMERICA
DESCRIPTION: Possesses layered, interlocking armor so tough it can surive being stepped on or run over.

Water Strider
Gerridae

COMMON LOCATION: FRESHWATER SURFACES WORLDWIDE
DESCRIPTION: Uses microscopic leg hairs to trap air, allowing it to walk and run across water without sinking.

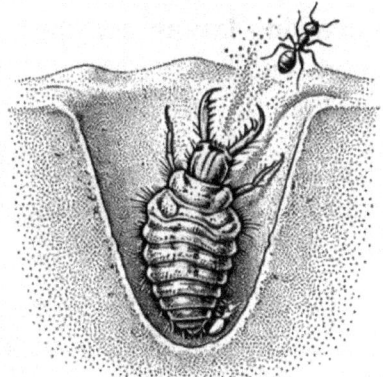

Antlion Larva
Myrmeleontidae

COMMON LOCATION: DRY, SANDY REGIING; WORLDWIDE
DESCRIPTION: Builds precision sand pits that cause prey to slide inward using gravity and physics.

The hummingbird hawk moth (found across Europe, Africa, and Asia) can hover like a hummingbird. Its fast wings allow it to sip nectar in midair.

Dragonfly nymphs shoot water jets from their rear end for propulsion. This jet burst helps them escape predators and strike prey with explosive force.

Termites (found in tropical and subtropical regions worldwide) build mounds with advanced ventilation. Their structures regulate heat and airflow using natural convection.

The bombardier beetle's explosion chamber can fire 500 blasts per second. Pulsing the reaction prevents the beetle from overheating itself.

Harvester ants have venom stronger than a cobra's relative to size. A few stings can incapacitate small animals.

The Malaysian stick insect (one of the world's largest insects) can reach lengths over 2 feet. Its leaflike camouflage fools predators completely.

Lacewing larvae use 'trash armor' for camouflage. They pile debris and insect parts on their backs to disguise themselves from predators.

The giant water bug (found in North America and Asia) can stab prey with a beak-like mouthpart. Its bite is powerful enough to take down fish and frogs.

Monarch butterflies navigate thousands of miles using the sun. An internal clock in their antennae helps them adjust for the sun's movement.

The atlas beetle (a rhinoceros beetle from Southeast Asia) has horns used for wrestling rivals. It can lift nearly 100 times its weight while fighting.

Earwigs fold enormous wings into tiny spaces using natural origami. Their wing-folding mechanics inspire micro-robot design.

Wood ants spray formic acid with pinpoint accuracy. They use this chemical weapon to overwhelm larger predators.

The Japanese giant hornet can fly over 25 mph. Its powerful wings generate unusually strong lift for its size.

Fog beetles (found in Namibia) harvest water from desert air. Their shells condense moisture so they can drink in environments with almost no rainfall.

Adult antlions can hover like mini drones. Their wing design allows tight midair control for catching flying insects.

Fleas (parasites found worldwide) can jump the equivalent of a human leaping over a 30-story building. Elastic pads in their legs store and release incredible energy.

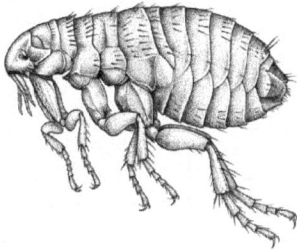

Flea

Siphonaptera

COMMON LOCATION: Worldwide.

Fleas can jump the equivalent of a human leaping over a 30-story building. Elastic pads in their legs store and release enormous energy, allowing powerful leaps despite their tiny size.

The velvet ant (a wingless wasp from desert regions) delivers one of the most painful stings of any insect. Its thick armor protects it during attacks.

The glasswing butterfly has almost transparent wings. Microscopic structures prevent light scattering, making the wings nearly invisible.

The green darner dragonfly migrates across North America in multi-generational waves. Each generation completes a different leg of the journey.

Green Darner Dragonfly
Anax junius

COMMON LOCATION:
North America.

DESCRIPTION:
Green darner dragonflies migrate thousands of miles across North America in multi-generational waves, with each generation completing a different leg of the journey—an epic relay race carried out entirely by instinct.

Camel spiders (large desert arachnids, not true spiders) can sprint over 10 mph. Their speed helps them outrun most small prey.

The bombardier beetle's spray turret can swivel. Tiny valves let it aim its boiling chemical jet left or right.

Ants can detect earthquakes hours before they happen. They sense changes in electromagnetic fields and ground vibrations.

MYTHS - BUSTED

Many people think insects are weak because they're small. Relative to their size, insects are some of the strongest creatures on Earth. Their muscles scale differently from ours, giving them super-strength nature never granted humans.

It's a common belief that insects with super strength are rare exceptions. In reality, extreme strength is normal across ants, beetles, and many small insects. Their exoskeleton and muscle design make powerful lifting a standard feature, not a rarity.

Some people think only large animals migrate long distances. Several insects-like monarch butterflies and green darner dragonflies-travel thousands of miles. Their long-range navigation rivals that of birds and marine mammals.

Many assume insects react instantly but cannot plan their movements. Dragonflies calculate interception paths before they strike, similar to guided missiles. Their precision is the result of neural systems tuned for predictive hunting.

People often believe beetles have simple, unremarkable shells. Some beetles-like the ironclad beetle-have armor tougher than many synthetic materials. Engineers study their shell structure for innovations in protective gear.

It's a common myth that insects are clumsy fliers. Dragonflies, robber flies, and hawk moths are elite aerial athletes that outperform most birds in maneuverability. Their wing designs allow acrobatics humans still struggle to replicate with drones.

Cricket
Gryllidae family

COMMON LOCATION: Worldwide.
Produces powerful sound by rubbing specialized wings together.

Monarch Butterfly
Danaus plexippus

COMMON LOCATION: North America.
Travels thousands of miles using inherited navigation instincts.

LEGENDS

LEGENDS

High in the Andes, travelers told of beetles that survived being stepped on by pack animals without injury. Early expedition journals described locals showing "spirit beetles" that could not be crushed, believed to house protective mountain spirits. Modern scientists recognize these descriptions as matching ironclad beetles, whose interlocking armor plates let them withstand enormous pressure-an ability so surprising it naturally evolved into legend.

In rural Japan, villagers spoke of a dragonfly they called the 'Wind Cutter' because it never missed its prey. Records from the Edo period mention hunters using the insect's flight as an omen of a successful season. Today we know dragonflies perform complex midair calculations to intercept prey, a precision so unreal it fueled centuries of myth about supernatural accuracy.

Deep in the deserts of Namibia, tribes told stories of a beetle that 'drank from the sky.' Oral histories described an insect standing on dune ridges at dawn as water formed on its back, believed to be a gift from ancestral spirits. Scientists later found fog beetles collecting moisture from morning fog using the bumps on their shells, a survival trick impressive enough to inspire mythology.

In Central America, farmers once feared the Hercules beetle as a forest guardian that could lift fallen branches with magical strength. Accounts from early Spanish explorers mention locals warning that the beetle could overturn storage baskets or tools. While exaggerated, these legends likely grew from real encounters with Hercules beetles using incredible strength to move debris while searching for mates.

In parts of Eastern Europe, elders told of a night-flying moth whose wings resembled ghostly faces. Folklore from the region spoke of these moths warding off evil spirits when they appeared near homes. Scientists now believe these stories reference hawk moths or atlas moths, whose startling wing patterns and silent flight made them perfect material for protective legends.

Along the Mediterranean coast, fishermen told tales of tiny creatures that could jump high enough to vanish into the sky. Early naturalists recorded these descriptions without understanding the animals behind them. Today we know springtails were the mysterious jumpers-soil-dwellers capable of launching themselves many times their body length, an ability so extreme it naturally shaped local myths.

FUN & WEIRD FACTS

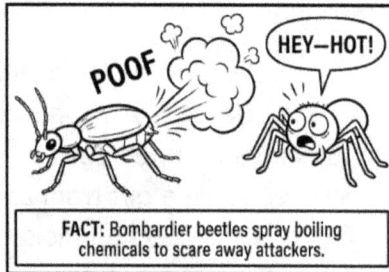

FACT: Stick insects use extreme camouflage to look exactly like twigs, fooling predators—and people.

FACT: Bombardier beetles spray boiling chemicals to scare away attackers.

FACT: Ants can lift objects many times their own body weight using powerful muscles and teamwork.

FACT: Cicadas make sound by vibrating special body plates, producing one of the loudest insect noises on Earth.

MIND-BLOWN™ Cartoons

Bug Smiles

A little boy and his dad are sitting at the dinner table when the little boy asks his dad, "Dad, are bugs good to eat?"

The dad responds, "Son, we're at the dinner table. We don't talk about things like that at dinner." The son says, "but Dad," only to have the father interrupt him and remind him that bugs are not appropriate dinner conversation.

Later that evening, after they have finished their meal, the dad asks, "Why were you asking about eating bugs at dinner? What do you need to know?" The son replies, "I don't need to know anything, anymore. There was a bug in your soup, but you ate it."

DID YOU KNOW?

Did you know? Dragonflies have a built-in neural "predictive calculator" that tracks moving prey by estimating where it will be next, not where it currently is. Their brains map the prey's flight path using rapid visual updates, allowing them to make interception turns with stunning accuracy. This system is so advanced that some robotics labs study dragonflies to improve drone-targeting algorithms.

Did you know? The incredible strength of ants comes from the way their tiny bodies scale. Their small size means their muscles take up more of their body volume and have less weight to move, allowing them to lift objects dozens of times heavier than themselves. Engineers study ant biomechanics to understand how lightweight robots might someday achieve similar force output.

Did you know? The explosive jump of a springtail happens because its furcula stores elastic energy far more efficiently than human muscles can. When released, this energy fires the creature upward in less than a millisecond. This mechanism is so effective that researchers compare it to biological nanotechnology used for ultra-fast motion.

Did you know? Bombardier beetles control their boiling chemical spray using a series of rapid-fire valves that open and close hundreds of times per second. This pulsing prevents the beetle from overheating itself while still delivering a powerful defensive burst. Scientists have modeled this system to explore safer ways of controlling micro-explosions in machinery.

Did you know? The shimmering colors of rainbow stag beetles are not pigments at all but microscopic structures that bend and scatter light. These layers act like natural prisms, creating iridescent colors that shift with angle. This form of structural color is so durable that the beetle keeps its shine long after it dies.

Did you know? Monarch butterflies navigate using a combination of the sun's position and an internal clock located in their antennae. This allows them to correct for changing solar angles during their migration across North America. No single monarch completes the entire journey, making it one of the most remarkable inherited navigation systems known.

Did you know? Termite mounds maintain stable temperatures using a complex system of vents and channels that operate like natural air conditioners. Hot air rises through central chimneys while cooler outside air circulates in, keeping the colony comfortable even when the outside temperature swings wildly. Architects have used termite-inspired ventilation in several real buildings.

Did you know? Robber flies rely on lightning-fast neural processing to calculate the exact angle needed to intercept prey midair. Their specialized eyes detect motion with exceptional precision, allowing them to grab insects traveling at high speed. This ability makes them one of the most formidable aerial predators among insects.

Did you know? Some beetles in extremely dry environments survive by harvesting water directly from air. Fog beetles in Namibia use bumps and valleys on their shells to condense moisture, which then rolls into their mouths. This adaptation is so efficient that scientists study it for inspiration in water-collection technology.

STORY MOMENT

STORY MOMENT

THE BEETLE THAT BROKE THE RULES

Dr. Lena Ortiz had handled tough insects before, but nothing prepared her for the ironclad beetle resting on her palm. It looked ordinary-slow, matte, unremarkable-yet the moment she applied gentle pressure with her forceps, the beetle didn't bend, crack, or even flinch. It simply tightened, as though its shell locked into place. Lena leaned closer. Something in its armor shifted, like puzzle pieces aligning in perfect formation.

The specimen had been collected from a rocky canyon in the Southwest, where locals joked that these beetles survived falling boulders. Lena had dismissed the rumors-until now. When she applied precise force meant to mimic a predator attack, sensors picked up impossible readings. No insect shell should withstand that much pressure. No arthropod should be able to "flex instead of break." Yet this one did, again and again, almost effortlessly.

It was late evening when she recorded her final notes. Outside, cicadas buzzed against the growing dusk. She turned off the overhead lab light, leaving only the soft glow of her desk lamp. The beetle shifted, crawling onto a small rock sample with the steady determination of something built to endure. Lena exhaled. For centuries, legends had formed around creatures that defied expectations. Tonight, watching the beetle settle quietly beneath the lamplight, she understood why. Some superpowers aren't loud-they're simply unbreakable.

FUN QUIZ

1. Which insect can survive being stepped on due to its interlocking armor plates?
    A) Tiger beetle
    B) Ironclad beetle
    C) Rainbow stag beetle
    D) Leafcutter ant

2. Dragonflies catch prey with exceptional accuracy because they:
    A) Glide silently
    B) Use predictive flight calculations
    C) Have infrared vision
    D) Produce ultrasonic clicks

3. Which insect creates light that is nearly 100% efficient?
    A) Firefly
    B) Cicada
    C) Housefly
    D) Praying mantis

4. Springtails launch themselves using what structure?
    A) Wing hinge
    B) Furcula
    C) Mandible spring
    D) Leg piston

5. Robber flies are known for their ability to:
    A) Build sand pits
    B) Catch prey midair with precision
    C) Produce ultrasonic songs
    D) Swim using trapped air bubbles

QUIZ ANSWERS

1. B) Ironclad beetle

2. B) Use predictive flight calculations

3. A) Firefly

4. B) Furcula

5. B) Catch prey midair with precision

6. False – they navigate using the sun and an internal clock.

7. True

8. B) Structural layers that bend light

9. B) Dragonfly nymph

10. B) Creating living bridges with their bodies

CHAPTER 5

5-Tiny Architects & Builders of the Impossible

Engineering, construction, group building, traps

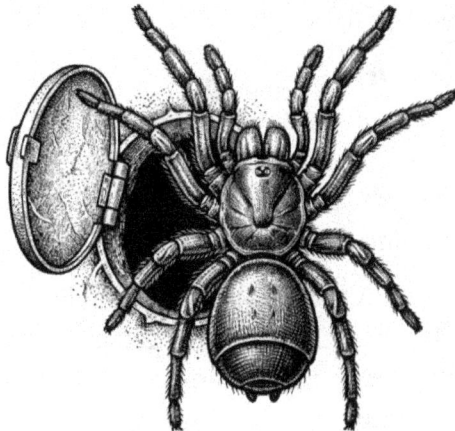

Trapdoor Spider
Cyclocosmia / Ctenizidae family

COMMON LOCATION: Worldwide. Builds camouflaged underground doors to ambush passing prey.

FUN & WEIRD FACTS

MIND-BLOWN
MOMENT

Termites build towering soil mounds that act like living skyscrapers. Underground ventilation tunnels regulate airflow and humidity perfectly. **Termites (social insects from tropical and subtropical regions)** engineered nature's first natural air-conditioning system, keeping their colonies cool all year long.

Honeybees construct flawless hexagons in their golden hives. This geometric brilliance stores the most honey using the least wax. **Honeybees (pollinating insects found worldwide)** inspired modern engineers designing lightweight and super-efficient structures.

Weaver ants sew leaves together with living silk. They pull leaf edges tight while using larvae as glue guns that emit silk strands. The tree-top fortresses of **weaver ants (arboreal ants from tropical Asia)** are waterproof masterpieces of teamwork.

Paper wasps chew wood pulp to build delicate, paper-like nests. Layer by layer, their homes harden into lightweight shells that resist wind and rain. The craftsmanship of **paper wasps (social wasps found in warm climates)** rivals handmade sculpture.

Sponge crabs cut perfect sponges into helmet-shaped armor. They carry their squishy shields everywhere for camouflage and protection. The partnership between **sponge crabs (small marine crustaceans from tropical reefs)** and their "living armor" is one of the ocean's cleverest disguises.

Trapdoor spiders engineer mossy burrows with secret entrances. Each silk-lined lid seals flush against the ground, making them invisible. **Trapdoor spiders (burrowing arachnids from forests worldwide)** wait quietly, then strike from their hidden homes with lightning speed.

Potter wasps sculpt perfect clay pots smaller than a grape. They fill each pot with paralyzed insects as food, seal it, and leave. The architectural genius of **potter wasps (solitary wasps found in warm regions)** can withstand even heavy rains.

Caddisfly larvae build armored suits from sand, shells, and silk. Each moving fortress grows as the insect grows. **Caddisfly larvae (aquatic insects from rivers and streams)** turn riverbeds into tiny construction sites bursting with creativity.

Leaf-rolling caterpillars fold leaves into spotless scrolls. Their silk stitches hold firm, making a fortress against predators. **Leaf-rolling caterpillars (moth larvae from tropical forests)** roll more precisely than a sushi chef's hand.

Dung beetles sculpt perfect spheres of dung faster than you can blink. They roll and bury them like treasure chests underground. By doing so, **dung beetles (beetles found worldwide)** recycle waste and improve the soil every single night.

Mud dauber wasps craft slender mud tubes with smooth walls. Inside, each chamber holds an egg and a carefully paralyzed spider for food. **Mud dauber wasps (solitary wasps found near water and cliffs)** are the world's tiniest clay masonry experts.

Army ants form living walls from their own bodies. Thousands link legs to build bridges, ceilings, and cradles. The teamwork of **army ants (predatory ants from rainforests)** creates architecture that breathes, flexes, and moves.

Leafcutter ants cultivate fungus farms underground. They trim and carry leaf fragments to nourish their crops, keeping them free of mold. **Leafcutter ants (tropical ants from Central and South America)** invented agriculture millions of years before humans.

Compass termites build ridged towers aligned north to south. This clever design limits sun exposure, keeping interiors cool. Compass termites (native to northern Australia) are natural landscape engineers-and accurate navigators.

Cathedral termites raise enormous spires taller than a person. Air shafts and chambers inside cool and ventilate the colony naturally. These **cathedral termites (African mound-building termites)** create skyscrapers that could teach architects a thing or two.

Cathedral Termites
Macrotermes spp.

COMMON LOCATION:
Sub-Saharan Africa.

DESCRIPTION:
Cathedral termites actively construct towering earthen mounds that rise far above the insects themselves, with visible vertical ridges and internal ventilation shafts. The termites remain large and clearly detailed in the foreground, while a section of the massive mound appears behind them a secondary element, emphasizing the scale of sophistication of these naturally climate-controlled skyscrapers.

Weaverbirds weave dangling nests from strands of grass. Each woven globe begins with a perfect knot and grows into a round, rainproof cradle. The artistry of **weaverbirds (songbirds from Africa and Asia)** turns entire trees into swinging cities.

Trap-building antlion larvae dig deadly sand pits shaped like funnels. Prey that slips in can't climb out-the sand acts like a slide. **Antlion larvae (predatory insect larvae found in sandy habitats)** prove that great architecture can be both beautiful and brutal.

Cellophane bees seal burrows with clear, waterproof linings that glisten like varnish. Scientists call it nature's "bioplastic." **Cellophane bees (ground-nesting bees found in temperate regions)** are tiny inventors of sustainable waterproofing.

Tree ants carve galleries inside hollow branches. They form rooms, tunnels, and even ventilation shafts. **Tree ants (arboreal ants from tropical forests)** rebuild fast when branches break, moving in and starting fresh.

Fairy ants build lacy soil towers that bend but never break. The secret is saliva, used to cement each grain perfectly in place. Storms can't topple **fairy ants (tiny tropical ants)** because their shimmering towers flex like living reeds.

Basket spiders weave silk baskets that hang above their own webs. These chambers act like panic rooms-safe hideouts from predators. **Basket spiders (orb-weaving spiders)** design air-light shelters that combine beauty with engineering.

Harvest mice weave perfect ball-shaped nests between tall grasses. The green fibers interlock so tightly that rain rolls off like wax. **Harvest mice (tiny rodents from grasslands of Europe and Asia)** create floating cradles in the wind.

Social spiders build team webs that blanket entire trees. Each worker maintains a section, while others rush to patch holes. **Social spiders (communal orb-weavers from tropical regions)** spin structures so massive they can catch birds.

Ant colonies become living machines during construction. **Ants (social insects found worldwide)** link arms to form ladders, bridges, and even shelters for the queen. Their living structures adapt, flow, and change with amazing precision.

Bagworm moth larvae carry mobile homes wherever they go. They build the cases from leaves, bark, and silk, adding new layers as they grow. **Bagworm moths (camouflaged caterpillars from forests)** create wearable architecture that doubles as armor.

Bagworm Moth Larva
Psychidae

COMMON LOCATION:
Forests and woodlands worldwide.

DESCRIPTION:
Bagworm moth larvae construct portable cases from leaves, bark, and silk, carrying these mobile homes wherever they go. As they grow, they add new layers, creating wearable architecture that provides camouflage and protection like living armor.

Coral polyps build breathtaking reefs that become entire ocean cities. Over centuries, each polyp's calcium skeleton fuses into coral stone. **Coral polyps (tiny marine animals worldwide)** shape coastlines while giving thousands of species a place to live.

Megachile bees cut perfect circles from leaves and petals. They line each brood cell like wallpaper and seal it tightly. **Megachile bees (leafcutter bees from gardens and meadows)** are nature's green architects.

Termite queens rule climate-controlled royal chambers deep underground. The walls regulate warmth and moisture perfectly for egg-laying. Without this precision, **termite queens (rulers of social termite colonies)** could not survive.

Beetle larvae carve hidden tunnels through dead wood. These pathways provide both shelter and food as the **beetle larvae (wood-boring insects from global forests)** recycle trees into new soil. Each burrow becomes a micro-labyrinth of life.

Compass ants orient nests eastward to catch the morning sun. The alignment warms them quickly after cold nights. **Compass ants (Australian desert ants)** prove even insects read nature's clockwork.

Twig-nest ants weave plant fibers around hollow twigs to form multi-room nests. These aerial shelters let **twig-nest ants (arboreal ants from tropical forests)** thrive high above the ground.

Mason bees mold clay walls to form small, sturdy chambers inside hollow stems. The sealed cells hold eggs and pollen for their larvae. **Mason bees (solitary bees found worldwide)** are peaceful construction pros essential to gardens.

Goblet wasps shape mud vases that look like miniature amphoras. They hide them under overhangs and fill them with food for their young. **Goblet wasps (tropical potter wasps)** blend artistry with parental care.

Hornet architects design multilayered paper nests cooled by wing-powered airflow. Groups of **hornets (large social wasps from Asia and Europe)** fan the hive constantly to keep it at the ideal temperature.

Driver ants tunnel through soil creating temporary underground megacities. Each colony moves and rebuilds every few days. **Driver ants (African army ants)** are nomadic builders whose architecture walks on six legs.

Sea anemones construct sand tubes as armor against shifting tides. The sticky mucus of **sea anemones (small marine predators)** binds grains together, creating flexible, living chimneys.

Tube-Building Sea Anemone
Ceriantharia (tube anemones)

COMMON LOCATION:
Shallow coastal waters worldwide.

DESCRIPTION:
Some sea anemones protect themselves by building flexible tubes made from sand and debris. Sticky mucus binds the grains together, forming living chimneys that anchor the animal against waves and shifting tides.

Trapdoor ants build hinged lids to seal off their tunnels at danger. When disturbed, **trapdoor ants (Australian subterranean ants)** slam the entrance shut in a click that echoes underground.

Fiddler crabs decorate burrow entrances with neat sand chimneys. Each male **fiddler crab (small coastal crab)** uses its sandy sculpture to attract mates like a proud architect.

Cuckoo wasps carve tiny doorways into other insects' nests. Their metallic shells reflect sunlight, keeping **cuckoo wasps (parasitic wasps found globally)** cool while they secretly remodel borrowed homes.

Cuckoo Wasp
Chrysididae

COMMON LOCATION:
Worldwide.

DESCRIPTION:
Cuckoo wasps carve tiny entrance openings into the nests of other insects and lay their eggs inside. Their hard, metallic shells reflect sunlight and help keep them cool while they quietly remodel borrowed homes without being detected.

Laceweaver larvae pile camouflage on their backs using twigs and dust. This portable disguise hides **laceweaver larvae (predatory lacewing larvae)** as they stalk prey among leaves.

Burying beetles dig underground nurseries lined with antiseptic secretions. The hidden chambers of **burying beetles (beetles found in forests worldwide)** keep their food and offspring pristine.

Velvet ants burrow through sand to invade other insects' nests. The thick armor of **velvet ants (wingless wasps from deserts)** lets them bulldoze their way into almost any home.

Crab spiders fold flower petals into perfect cones for

camouflage. Waiting inside, **crab spiders (ambush hunters from gardens and meadows)** vanish among blooms until an unsuspecting prey lands within reach.

Cicada killers excavate deep tunnels with branching chambers for each paralyzed prey. The precision of **cicada killers (giant wasps from North America)** makes their nests both deadly and efficient.

Snail architects patch dying coral with cemented sand grains. Over time, the repairs made by **tiny reef snails (marine snails from tropical coasts)** rebuild coral structures after storms.

Tower termites erect vertical chimneys that vent hot air out and draw cool air in. The elegant design of **tower termites (African savanna termites)** shows that good climate control can exist without electricity.

Pseudoscorpions weave silk chambers under bark or stones. Small but fierce, each **pseudoscorpion (tiny predatory arachnid)** crafts a perfect micro-shelter to hide from danger.

Sand masonry worms glue seashells into shimmering tubes that rise from the sea floor. The coral-like castles of **sand masonry worms (marine worms found worldwide)** sparkle like underwater cathedrals.

Fire ants form waterproof rafts during floods by linking limbs together. Thousands of **fire ants (social insects from the Americas)** float for weeks without sinking-a true miracle of biological engineering.

Glow-worm larvae dangle glowing silk to lure prey in the dark. Their sticky threads form a living chandelier. **Glow-worm larvae (bioluminescent insects in caves and forests)** create some of nature's most magical architecture.

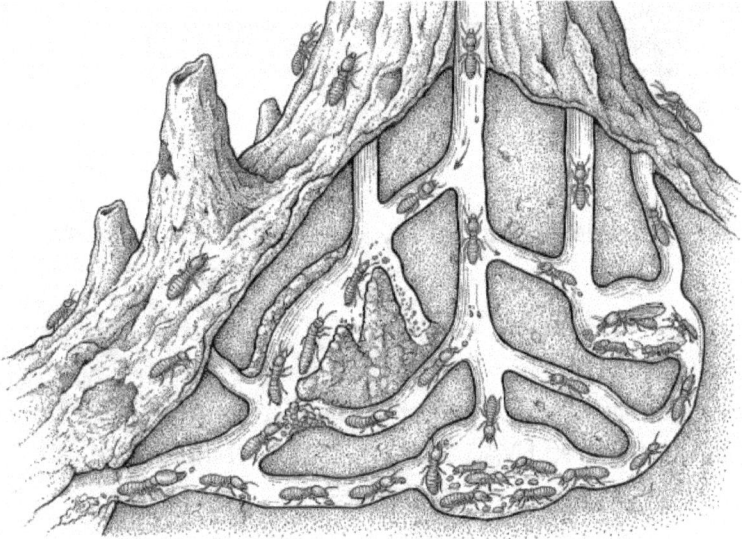

Mound-Building Termite *Macrotermes spp.*

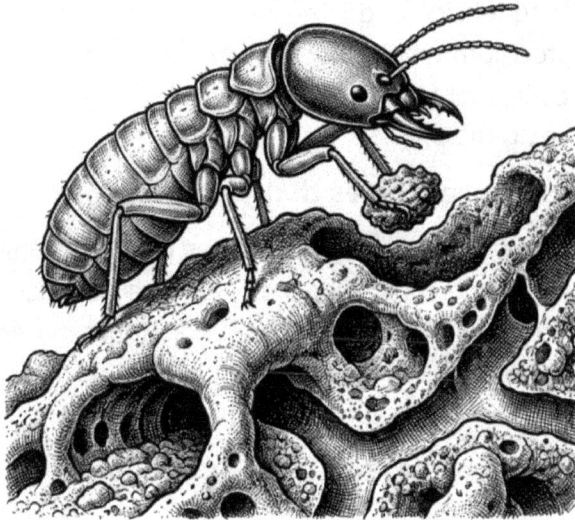

Termite
Isoptera

COMMON LOCATION: Tropical and subtropical regions worldwide.
Builds complex mounds and tunnel systems that regulate airflow, temperature, and moisture.

MYTHS - BUSTED

Termite mounds are just dirty piles built by mindless bugs.

Reality: Termite mounds are climate-control machines built by blind builders who never see the final structure. They adjust airflow, humidity, and oxygen gradients with earthen vent stacks that rival engineered HVAC systems. Inside, temperatures stay more stable than many human homes.

Spider webs are random sticky messes tossed between branches.

Reality: Orb-weaver spiders tune every strand like a musical instrument, setting tension and vibration frequency with mathematical precision. When prey hits the web, the spider reads the signal like coded data, identifying size, species, and direction without ever seeing the prey.

Ant tunnels are random holes that just happen to connect.

Reality: Ant colonies build underground cities using air pressure, carbon dioxide gradients, and soil density as invisible blueprints. Their tunnels form optimized transport networks that mirror advanced human-engineered systems long before we discovered similar principles.

Paper wasps make nests out of spit and luck.

Reality: Paper wasps chew wood into microscopic fibers and blend it with enzymes to form a lightweight structural composite stronger, gram for gram, than some human-made materials. Every chamber is placed with astonishing accuracy, expanding the nest like a living skyscraper.

Trapdoor spiders just hide in holes.

Reality: Trapdoor spiders engineer perfectly hinged, camouflaged doors balanced with soil, moss, and silk. The door acts like a natural stealth trap with a pressure-sensitive alarm system, sealing so tightly that predators strike past it without realizing a spider is beneath.

Caddisfly cases are just sticks glued together.

Reality: Caddisfly larvae are underwater architects that hand-select stones, sand, bone fragments, and shell pieces to build armored cases shaped by water flow physics. Their designs serve as protective shields, camouflage, and ballast-each species crafting a signature architectural style.

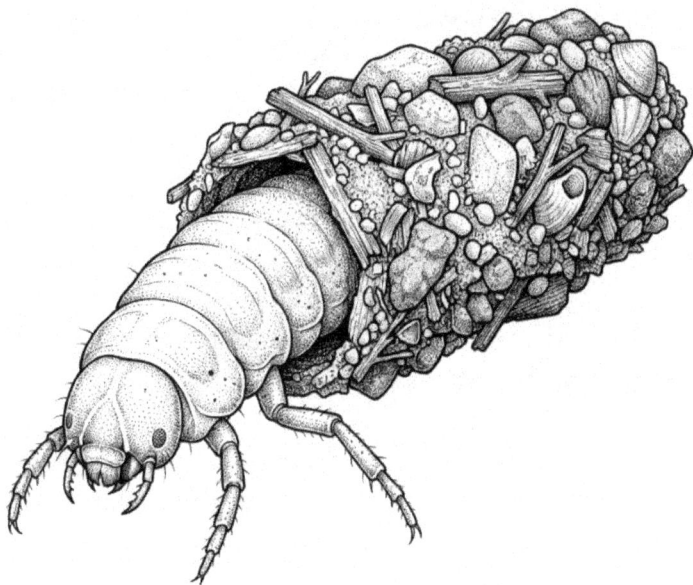

Caddisfly Larva
Trichoptera order

COMMON LOCATION: Freshwater streams worldwide.
Builds a protective case from surrounding materials using silk as cement.

LEGENDS

LEGENDS

The Burrow That Breathed

Old miners in the Australian outback once told stories of a massive **termite mound** that seemed alive in a way no one could explain. On still nights, they said you could hear air rushing through it in long, pulsing breaths - as if something huge was sleeping inside. When a drought hit and animals began vanishing, locals claimed the **mound** grew taller overnight, feeding on anything that wandered too close. No one ever proved it, but explorers reported the ground around it was warm, even when the desert was freezing. Some swear the **mound** wasn't a home - it was a living **creature** wearing termites like skin. Wow.

The Spider Gate of Katoa Forest

In the remote Katoa Forest, travelers spoke of a circular **web** so large it stretched across an entire ravine - a **web** so perfect it looked woven by many hands, not eight legs. Locals warned never to pass underneath, because animals that entered the ravine were often found days later, wrapped in silk hammocks suspended from the trees. Stranger still, the silk bundles always hung in a precise spiral, like part of some enormous design. Some hunters believed the spider wasn't trapping prey... it was building something from them. And when one bundle twitched on its own, villagers simply said, 'That's why we never walk beneath the Gate. 'Eww'.

The Endless Ant City

In rural Argentina, farmers told tales of an **ant** colony that stretched so far underground no one ever found its end. Horses sometimes crashed through thin soil ceilings and into vast tunnels, some wide enough for a person to walk through. One explorer followed the tunnels for hours before feeling vibrations beneath his boots, like footsteps coming from every direction. He insisted the **ants** weren't running toward something - they were guarding something deeper. Whatever he heard echoing in that darkness made him flee whispering, 'The **city** isn't for the ants... it's for something else.' Wow.

The Caddis Stone Coffin

In a mountain stream in **Romania,** villagers spoke of a drifting **stone coffin** that appeared every spring. They said it was built by a monstrous **caddisfly larva** that chose fragments of bone washed down from old battlefields instead of pebbles. Travelers claimed the case glowed faintly at night, as if something inside was moving. One fisherman hooked it once - and when it surfaced, the case cracked open, releasing a swarm of wriggling larvae that clung to him like living gravel. He ran home screaming that the river was building soldiers of its own. Eww.

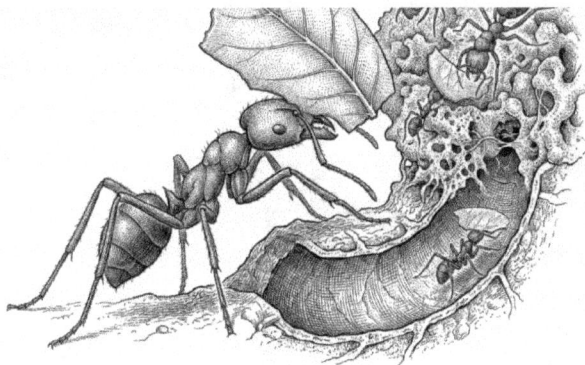

Leafcutter Ant
Atta spp.
COMMON LOCATION: Central and South America.
Builds underground fungus farms using harvested leaves.

DID YOU KNOW?

🪳 **Did you know?** Some species of **termites** build internal mud elevators that lift moisture upward through capillary action, keeping their underground fungal gardens alive during drought. This passive water-transport system is so efficient that engineers study it as a model for climate-resilient architecture.

🪳 **Did you know?** Certain trapdoor spiders weave delicate tension fibers inside their burrows that function like seismic sensors. When an insect walks across the soil above, the spider interprets the vibration signature to identify size, distance, and direction with astonishing precision.

🪳 **Did you know?** Colonies of weaver ants use living chains of workers to measure distances before construction begins. These chains stretch across gaps like flexible tape measures, allowing ants to calculate support points and load distribution long before a leaf is stitched into place.

🪳 **Did you know?** Some caddisfly larvae purposely collect materials with different densities-glassy grains, bone fragments, heavy stones-to fine-tune the weight of their portable armor. By adjusting the mix, they control buoyancy and current stability, turning each case into a personalized underwater engineering project.

🪳 **Did you know?** Certain species of paper wasps maintain their nests by vibrating their bodies against the combs, using heat and rhythmic motion to regulate humidity. This micro-climate control protects developing larvae from fungal infection and temperature swings. ants include waste-processing chambers where the colony composts discarded fungus. These chambers generate heat

STORY MOMENT

The Mound That Adjusted Itself

The forest floor hums as you kneel beside the strange mound, its surface trembling like something breathing beneath it. A thin column of warm air rises from a slit in the soil, carrying the scent of damp earth and something faintly sweet, almost metallic. When you brush away loose debris, you notice the tunnel walls aren't random dirt-they're layered, ribbed, sculpted, as if hundreds of tiny architects carved them with absolute purpose. A flicker of movement catches your eye, and for a second the whole chamber seems to pulse, the tunnels shifting their airflow as if adjusting to your presence. You lean closer, and the warm breath becomes a sudden cool draft, like the mound is testing you. Then the soil under your hand subtly firms, as if something inside has braced itself. You don't know whether to crawl backward or stay perfectly still... because for the first time, you're not sure if you discovered the colony-or if the colony just discovered you.

A soft clicking rises from deep inside the mound, almost like distant gears turning. The airflow steadies, the temperature shifts again, and something unseen rearranges the tunnels in complete silence. Whatever is beneath you isn't reacting by instinct-it's responding.

FUN QUIZ

1. What surprising function does a termite mound perform that makes it more like a natural machine than a pile of dirt?
Explain the environmental system it regulates.

2. True or False: A trapdoor spider can identify prey size and direction without seeing it.
Write "true" or "false."

3. Which behavior allows weaver ants to measure construction distances before building their nests?
A. Forming living chains
B. Dragging leaves in straight lines
C. Using scent trails as rulers

4. Why do caddisfly larvae select materials of different weights when building their underwater cases?
Explain how this affects movement and stability.

5. True or False: Leafcutter ant colonies include specialized waste chambers that help protect the colony from toxins.
Write "true" or "false."

6. Which process allows paper wasps to maintain the proper temperature and humidity inside their nests?
A. Fanning the nest with their wings
B. Vibrating their bodies against the combs
C. Dropping water into the chambers

7. Why might a mound appear to breathe or shift airflow when someone approaches it?
Describe the structural feature responsible.

QUIZ ANSWERS

1. It regulates airflow, humidity, and temperature like a natural climate-control system.

2. True.

3. A. Forming living chains.

4. To adjust buoyancy and stability in moving water.

5. True.

6. B. Vibrating their bodies against the combs.

7. Because its tunnels create a passive airflow system that changes with pressure and movement.

8. Size, direction, and type of prey from vibration patterns.

Honeybee *Apis mellifera*
COMMON LOCATION: Worldwide. Builds hexagonal wax combs used for food storage and raising young.

6-Creatures of Light & Fire

Bioluminescence, light communication, glowing traps

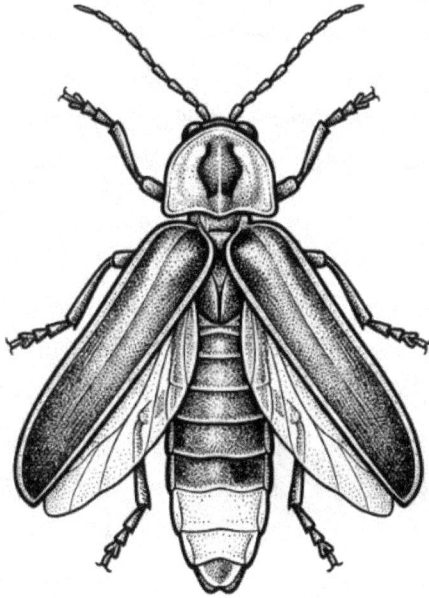

Firefly / Lightning Bug
Photinus spp.

COMMON LOCATION: Worldwide. Produces light through
a chemical reaction to communicate and attract mates.

FUN & WEIRD FACTS

Some species of railroad worms glow in two colors at once-green down the body and blazing ruby-red from the head. The dual glow of **railroad worms** (bioluminescent beetle larvae from the Americas) is so unusual that few land animals can even see the red wavelength, making them nearly invisible to predators.

The blue ghost millipede emits an eerie cyan halo that outlines its entire body like a moving hologram. Scientists have no clear reason for the glow because **blue ghost millipedes** (bioluminescent millipedes from the southern U.S.) already pack toxins so potent that warnings seem unnecessary.

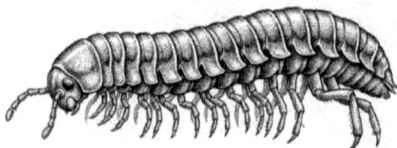

BLUE GHOST MILLIPEDE
Biomillipeda lucens

The blue ghost **millipede** emits an eerie cyan halo that outlines its entire body like a moving hologram. Scientists have no clear reason for the glow because blue ghost (bioluminescent millipedes from the southern U.S.) already pack toxins so potent that warnings seem unnecessary.

The larvae of the New Zealand glowworm suspend hundreds of glowing silk strings from cave ceilings. The strange galaxy-like shimmer lures prey upward toward sticky light. **Glowworm larvae** (fungus gnats from New Zealand caves) create the southern hemisphere's most famous natural light show.

The bombardier beetle blasts boiling chemicals from a built-in cannon. Its internal reaction chamber fires in rapid pulses that can hit targets behind it. **Bombardier beetles** (ground beetles found worldwide) are walking chemistry experiments with perfect aim.

The firebrat thrives beside ovens and hot boilers where other insects would bake alive. **Firebrats** (heat-tolerant silverfish from human dwellings) survive temperatures above 120 °F thanks to enzymes engineered for heat. They're true house-made fire specialists.

Some Central American click beetles carry head lanterns bright enough to read by. Early travelers used handfuls of **click beetles** (bioluminescent beetles from tropical forests) as living flashlights inside tents-a glowing handful of history.

The Arizona luminous millipede glows brighter when frightened. Cyanide in its skin releases light and warning color at once. The dangerous beauty of **Arizona luminous millipedes** (toxic desert millipedes from the U.S. Southwest) lights up dark canyon floors.

The tropical railroad click larva flashes in sequence like a miniature moving train. Each pulse travels down the body in perfect rhythm. **Railroad click larvae** (bioluminescent beetles from Central America) may use this light choreography to signal one another under leaves.

Some paper-lantern glow beetles shine nonstop from dusk to dawn. The constant green glow helps **glow beetles** (tropical beetles from Southeast Asia) disappear into moonlit foliage instead of pulsing like fireflies.

The hot-spring springtail skims across steaming

geothermal pools that would scald human skin. **Hot-spring springtails** (tiny arthropods from volcanic geyser fields) possess proteins that stay stable near boiling point-true jumpers of the inferno.

Paper Lantern Glow Beetle
Pyrocoelia spp.

COMMON LOCATION:
Southeast Asia.

DESCRIPTION:
Some glow beetles shine with a steady green light from dusk to dawn instead of flashing like fireflies. This constant glow helps them blend into moonlit foliage, making their light a form of camouflage rather than a signal.

The ashy ember moth lays eggs only in warm fire ash after wildfires. Residual heat incubates the brood of **ashy ember moths** (post-fire moths of Western North America) whose larvae feed on burned organic debris no other species can stomach.

The red lantern gnat shines with deep crimson light, one of nature's rarest hues. The pulsing glow from **red lantern gnats** (bioluminescent midges from tropical Asia) turns forest floors into fields of flickering hearts.

The volcano silverfish survives inches from steaming vents. Its waxy body armor keeps **volcano silverfish** (heat-adapted insects from volcanic slopes) safe from blasts of vapor that would melt other species instantly.

The fire-seeking beetle Melanophila flies toward infernos it can sense 50 miles away. Infrared sensors guide **Melanophila beetles** (metallic wood-boring beetles found worldwide) straight to charred trees where their larvae feast in heat-sterilized wood.

The cave halo cricket sparkles softly in fungus-lit darkness. Reflective panels on **cave halo crickets** (cave-dwelling insects from Southeast Asia) bounce ambient glow into moving dots of false light that fool predators into striking empty air.

The blue flash springtail erupts in a burst of color whenever it leaps. Microscopic prisms scatter light into instant iridescence, disorienting hunters. **Blue flash springtails** (soil springtails from South America) vanish again before reality catches up.

The ember antlion warms sand inside its pit traps with solar-absorbing skin. Heated grains slow the scrambling of prey. **Ember antlions** (heat-loving insect larvae from deserts) make blazing arenas no captive can escape.

The desert phosphor beetle stores daylight energy like a battery. At night **desert phosphor beetles** (nocturnal beetles from arid North Africa) leak it as pale blue shimmer, leaving glowing trails across dunes.

The luminous rove beetle glimmers with warning light when disturbed. Abdominal glands release chemicals that glow soft green. **Luminous rove beetles** (beetles from moist forests) use bioluminescence as both defense and identity badge.

The forest spark moth scatters sunlight like embers when flying through beams. Hidden pigments on **forest spark moths** (day-active moths from tropical forests) ignite into fiery flashes that vanish in shade.

The thermal ridge weevil perches on black volcanic rock hotter than 130 °F. **Thermal ridge weevils** (weevils from volcanic deserts) stay active thanks to reptile-grade heat-tolerant proteins that keep muscles from cooking.

Thermal Ridge Weevil
Hypothenemus spp.

COMMON LOCATION:
Volcanic desert regions.

DESCRIPTION:
Thermal ridge weevils can remain active on black volcanic rock heated above 130°F. Specialized heat-tolerant proteins protect their muscles and enzymes from breaking down, allowing movement where most insects would overheat.

The ghost-tunnel glow larva drills spirals through damp soil that faintly shine with its glowing waste. When neighboring tunnels twinkle, **ghost-tunnel glow larvae** (bioluminescent beetle larvae) sense one another without ever meeting.

The sunstone leafhopper flashes metallic orange like flickering coals. These light-scattering plates make **sunstone leafhoppers** (tropical sap-feeding insects) seem to pulse as they move, confusing predators that hunt motion.

The firegrass katydid hides in tinder-dry grass that easily burns. Its yellow-green skin matches the filtered light of smoke. **Firegrass katydids** (grasshoppers from drought-prone savannas) may have evolved camouflage for glowing firelight.

The dune flare beetle uses searing desert heat to crack closed seed pods. Once they split, **dune flare beetles** (desert beetles from North Africa) feast before less-tolerant competitors arrive.

The glossy night cicada shines dark blue under moonlight despite its black shell. The reflective cuticle of **glossy night cicadas** (nocturnal cicadas from East Asia) scrambles depth perception for bats that chase them.

The luminous mud wasp larvae glow faintly from underground brood cells. The dim light guides adult **mud wasps** (riverbank wasps from tropical regions) back to their nests at night-then fades forever after pupation.

The starlight lacewing larvae sparkle with micro-mirrors on their backs. Moving through foliage, **starlight lacewing larvae** (predatory insect larvae from rainforests) resemble drifting constellations of dust and light.

The ghost ember millipede gleams only at its leg joints. To predators, **ghost ember millipedes** (bioluminescent millipedes from tropical forests) seem like clusters of hovering dots instead of one body-an optical illusion defense.

Some lava-tube planthoppers live entire lives in sunless caves. Guided by warmth gradients, **lava-tube planthoppers** (cave insects from volcanic islands) behave more like blind deep-sea creatures than daylight insects.

The alpine glow midge endures freezing nights near snowmelt streams. Stored anti-freeze compounds also fluoresce blue under UV light. **Alpine glow midges** (mountain midges from high-altitude Asia) protect their cells while sparkling in sunlight.

The ember termite-hunter beetle raids smoldering nests just after fires. Smoke masks the scent of **ember termite-hunter beetles** (predatory beetles from African savannas) long enough for them to steal larvae from weakened colonies.

Ember Termite Hunter Beetle
Thermophilous Carabidae
(termite-hunting ground beetles)

COMMON LOCATION:
African savannas and fire-prone grasslands.

DESCRIPTION:
Ember termite hunter beetles raid termite nests just after wildfires, using lingering smoke to mask their scent. This brief window lets them slip into weakened colonies and steal larvae before defenses recover.

The bioluminescent water boatman glows beneath the surface when disturbed. Expanding ripples amplify its light ring outward. **Bioluminescent water boatmen** (aquatic insects from freshwater pools) use light like underwater radar.

The fireline moth appears only in midday heat shimmering above burned ground. The vibration of air hides **fireline moths** (heat-adapted moths from arid regions) in living mirages that confuse birds.

The iridescent flare mantis flicks sunlight from its flashing forelegs. The burst blinds prey for an instant, giving **flare mantises** (mantids from tropical plains) perfect strike timing.

The silver-vein stick insect glides with moonlit wings that shine like threads of flame. When at rest, the transparency of **silver-vein stick insects** (nocturnal leaf insects from Asia) turns eerie shimmer to still shadow.

The thermal sand hopper leaps across dunes so hot they could fry an egg. Special pads under the feet of **thermal sand hoppers** (amphipods from desert coasts) spread heat faster than it can burn them.

The afterglow beetle releases warmth rather than visible light. At night **afterglow beetles** (tropical beetles from South America) radiate faint infrared heat detectable only with thermal cameras.

The fire-edge crane fly thrives along fresh burn scars where microbes bloom in charcoal soil. The larvae of **fire-edge crane flies** (insects from post-fire grasslands) feed on this living soot others can't digest.

Fire-Edge Crane Fly
Tipulidae

COMMON LOCATION: Post-fire grasslands and burned forest edges.

DESCRIPTION: Fire-edge crane flies thrive along fresh burn scars where microbes rapidly colonize charcoal-rich soil. Their larvae feed on this living soot, digesting resources that most other insects cannot use.

The shimmer ant mimic gleams like molten metal as it races across rock. Heat-proof scales make **shimmer ant mimics** (beetles that imitate ants) appear as wandering sparks, confusing true ants and predators alike.

Certain luminous barklice glow only while feeding on radiant lichens. When the glowing growths fade, so does **luminous barklice** (bioluminescent insects from humid forests) -a seasonal light that turns off with its meal.

The sun-glint aphid flashes sunlight in tiny bursts as it turns on stems. Those mirror-like flares make **sun-glint aphids** (sap-sucking insects from meadows) mimic larger wings, tricking hungry birds.

The copper flare beetle absorbs early-morning rays faster than its rivals. Within minutes **copper flare beetles** (metallic beetles from deserts) are 25 °F warmer than air, starting the day before others can move.

The moon-shard lacewing shines with powdered silver-blue wings. Moonlight refracts through them into cold flame as **moon-shard lacewings** (night lacewings from tropical Asia) glide between trees like drifting ribbons.

The glarewing meadow fly scatters sunlight into shifting blazes mid-flight. When it lands, the brilliance vanishes completely. **Glarewing meadow flies** (grassland flies from open plains) cloak themselves in disappearing light.

The sunburst katydid nymph glows brighter in strong sunlight. The blaze-yellow patterns of **sunburst katydids** (grasshoppers from equatorial forests) warn predators that brightness can mean danger.

The emberline assassin bug cooks prey from the inside out. Injected enzymes raise internal temperature until tissue liquefies. For **emberline assassin bugs** (tropical predatory insects) , dinner is literally hot.

Fire Beetle
Melanophila spp.

COMMON LOCATION: Forested regions worldwide. Detects infrared heat from fires to locate freshly burned trees.

Weevil
Curculionidae family

COMMON LOCATION: Worldwide
Known for its long snout used to drill into plants.

Bug Smiles

- What's a mantis's favorite movie? *The Good, the Bad, and the Mantis*!

- Why did the mantis get an award? For her outstanding *praying* in his field!

- How do you know a mantis is happy? He's *praying* for a good time!

- Why was the mantis so popular? He had great *forearms* for high-fives!

- What do you call a mantis who loves music? A *mantis*-ician!

Glowworm
Lampyridae family

COMMON LOCATION: Worldwide.
Produces cold light to attract prey or mates.

Glowworm Key Facts

- **Bioluminescence:** Light attracts mates (females) and prey (larvae).

- **Hunting:** Larvae build sticky silk snares, like spiderwebs, to catch small flying insects.

- **Habitat:** Damp, dark places – caves, forests, streamsides, grasslands.

- **Diet:** Larvae eat slugs, snails, midges, and flies; adults often don't eat.

- **Life Cycle:** Short-lived adults (days), focused on mating; larvae live longer.

- **Hunger Glow:** A hungry glowworm (larva or female) glows brighter.

MYTHS - BUSTED

MYTHS

Some people believe glowing insects generate heat like tiny embers burning inside them. In truth, **bioluminescent insects** (species such as click beetles and glowworms) produce one of nature's most efficient chemical reactions, releasing light without noticeable warmth. Their radiance is literally "cold light," so even the brightest glow stays cool to the touch.

Many think fire-loving beetles swoop toward flames for comfort or fascination, but the truth is strategy, not curiosity. **Melanophila beetles** (heat-seeking wood-borers from fire zones worldwide) detect infrared radiation that points to freshly burned trees where they lay their eggs. To them, wildfire ash is real estate, not a bonfire.

Hikers sometimes assume glowing millipedes are radioactive because their light seems to leak from nowhere. The effect actually comes from special enzymes reacting with oxygen near cyanide compounds in the shell. For **luminous millipedes** (bioluminescent millipedes from tropical regions) , the glow is a toxic warning, not nuclear magic.

Old forest tales claim fireflies brighten when storms approach because they can sense lightning. While flash frequency can rise as humidity climbs, **fireflies** (bioluminescent beetles found worldwide) use that pattern to flirt, not forecast weather. Stormy evenings simply make the perfect romantic backdrop.

A common rumor says luminous insects glow stronger when you warm them in your hand. Their light may change, but it's stress, not temperature, that triggers it. **Glowing beetles and worms** (cold-light insects from forests and fields) rely on oxygen and enzymes for power-no heat charging required.

Bug Smiles

- What does a caterpillar do on New Years Day?- Turns over a new leaf

- What do insects learn at school? – Mothmatics

- What do you call an insect on the moon? – lunar tick.

- Which insect is smarter than talking parrot? – spelling bee.

- What letter can hurt you if it gets too close? – B (bee)

- What did one firefly say to the other? – Got to glow now!

- Why are spiders like tops? – They are always spinning!

LEGENDS

LEGENDS

Long ago, hikers in the Blue Ridge Mountains whispered about the "blue ghost trail," a place where glowing millipedes formed drifting lines of light along the forest floor. Some said the lines moved with a purpose-curving, pausing, then flowing again like soft handwriting across the leaves. One night a ranger followed the glow, only to find the trail ending abruptly at a cold, empty patch of ground. The millipedes were gone, but the faint blue shimmer remained for several seconds, as if the forest itself had kept the last stroke.

Deep inside a volcanic cave in Hawai'i, researchers reported a cluster of heat-loving planthoppers behaving unlike any they'd seen. The insects crawled in slow, coordinated spirals around a warm vent, their pale bodies reflecting faint light from microbial mats. When the scientists approached, the spiral froze-every insect motionless-before breaking apart in perfect silence. Thermal sensors showed a temperature drop around the vent, though nothing in the cave could have caused it.

Campers along the Rio Negro sometimes speak of the "ember beetles," insects that seem to glow like dying coals scattered across the sand. Locals claim the beetles brighten when a large animal approaches, as if sensing something in the dark. One fisherman swears the beach lit up around him moments before a jaguar emerged from the brush. Whether the beetles reacted to vibrations, heat, or something else, no one has offered a solid explanation.

In parts of coastal Japan, night walkers describe seeing ribbon-like streaks of pale blue light drifting between cedar trees. These sightings often come after rainfall, when lacewings glide silently through the mist. Some witnesses report that the glowing streaks split in two, loop behind them, or appear suddenly at eye level. Photographers who've tried to capture the phenomenon claim their cameras malfunction or record only blurred arcs of light.

High in the Andes, villagers tell of a tiny insect they call "the warm whisper," a midge said to hum softly before glowing under starlight. Travelers have described feeling sudden warmth on their skin just before noticing the faint green shimmer of dozens of midges rising from the grass. The glow fades almost instantly, but the warmth lingers-sometimes for minutes. Scientists studying the region's UV-reflective insects say they can't rule out a temperature anomaly, though they've never witnessed the event themselves.

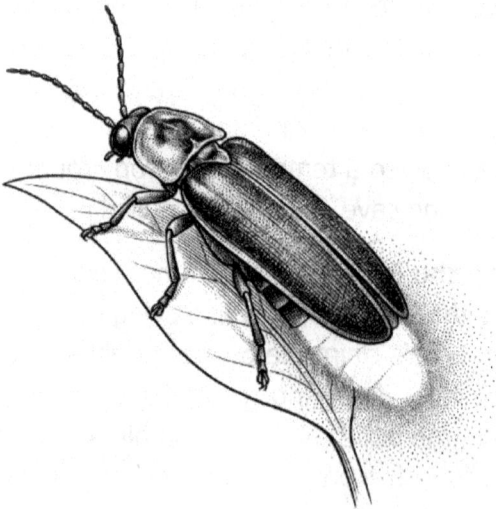

Firefly
Lampyridae family

COMMON LOCATION: Worldwide. Produces cold
light using bioluminescent chemistry.

DID YOU KNOW?

Did you know? Some species of **fire-loving beetles** carry specialized infrared sensors in their thorax that can detect the heat signature of a wildfire long before smoke reaches the horizon. These receptors are so sensitive that the insects can pinpoint burning trees from miles away, navigating with a heat-vision system rivaling some artificial sensors used in environmental monitoring.

Did you know? The cool glow produced by **bioluminescent millipedes** comes from a chemical reaction that is nearly 100 percent energy-efficient. While most chemical processes release heat as waste, these millipedes convert almost all of that energy directly into light, which is why their glow stays cold even when extremely bright.

Did you know? Certain **cave glow gnats** regulate the brightness of their light by adjusting the flow of oxygen to their luminous organs. When food is scarce, they dim their glow to conserve energy; when prey approaches, the glow intensifies, turning the cave ceiling into a living hunting trap.

Did you know? Some desert insects, like the **ember katydid**, reflect sunlight using microscopic scales that create fire-like flashes. These flashes may confuse predators that rely on motion detection, making the insect appear to flicker like a drifting spark rather than a solid target.

Did you know? The chemical blast produced by the **bombardier beetle** is triggered by mixing two normally harmless chemicals inside a reinforced chamber. Only when the beetle contracts its internal valves do these chemicals combine, creating a rapid, heat-generating reaction that boils instantly and bursts outward in a series of explosive pulses.

Did you know? Forest researchers have found that certain species of **glow wing lacewings** amplify faint moonlight through translucent veins in their wings. This creates the illusion of soft blue streaks drifting through the trees-an effect that may help individuals stay together while navigating after dark.

Glow-Wing Lacewing
Chrysopidae

COMMON LOCATION:
Forests and woodlands worldwide.

DESCRIPTION:
Lacewings do not produce light, but their translucent wings and fine vein patterns can reflect and scatter faint moonlight. As they fly through dark foliage, this creates the illusion of soft drifting streaks, which may help individuals stay visually grouped while navigating at night.

STORY
MOMENT

STORY MOMENT

The Valley That Breathed Fireflies

You notice it just after sunset-a faint shimmer seeping through the stone like trapped lightning. At first it looks like the day clinging stubbornly to the rocks, but the air begins to pulse. Thin green veins of light thread through the cracks, blinking in sequence, and before your eyes tiny winged forms drift upward, each one flaring and fading until the valley ripples like a low tide of fire. The light moves in waves, chasing itself from ridge to ridge as though the ground beneath you is breathing.

Heat brushes your skin, subtle and rhythmic, lifting with every pulse. The insects spiral together in tightening rings, the glow climbing toward the dark sky where it softens to blue. Beneath the hum of wings, a vibration begins-steady, almost musical-a sound you feel more than hear. The illumination swells until the walls of the valley shine with reflected emerald flame, brighter than moonlight yet cool enough to touch. For a moment, you stand inside a breathing constellation, perfectly timed, perfectly alive.

Then, as suddenly, the pattern breaks. The light fractures into a thousand sparks, scattering across the stone like the last embers of a fading campfire. The hum drops to silence. In its place lingers only warmth on your hands and a faint afterglow etched in the cracks-evidence that, deep in the valley's quiet heart, the creatures of light and fire are still awake.

MIND-BLOWN™ Cartoons

FUN QUIZ

1. What makes the light produced by glowing millipedes "cold," even when it appears bright?
Explain what happens to the energy in the reaction.

2. True or False: Fire-loving beetles fly toward flames because they enjoy heat.
Write "true" or "false."

3. Which structure allows certain beetles to sense distant wildfires?
A. Infrared receptors
B. Antennae with heat sacs
C. Temperature-sensing compound eyes

4. Why do some cave glow gnats brighten their glow when prey approaches?
Explain the advantage this gives them.

5. True or False: Desert insects that shimmer like fire have tiny internal flames.
Write "true" or "false."

6. What allows the bombardier beetle to produce boiling-hot chemical bursts?
Describe how the chemicals behave inside its reaction chamber.

7. Which adaptation helps lacewings appear as drifting blue streaks at night?
A. Reflective scale patches
B. Bioluminescent glands
C. Translucent wing veins that amplify moonlight

QUIZ ANSWERS

1. Because nearly all energy becomes light instead of heat; the reaction is extremely efficient.

2. False. They approach wildfires because infrared signals fresh habitat for their larvae.

3. A. Infrared receptors.

4. It intensifies the lure, drawing prey toward their sticky threads in dark caves.

5. False. Their "fire-like" glow is created by reflected or scattered light, not combustion.

6. It mixes two stable chemicals inside a reinforced chamber, triggering an instant heat-producing reaction that ejects boiling spray.

7. C. Translucent wing veins that amplify moonlight.

CHAPTER 7

7-Micro-Jungle Monsters

Soil, leaf-litter predators, unseen tiny species

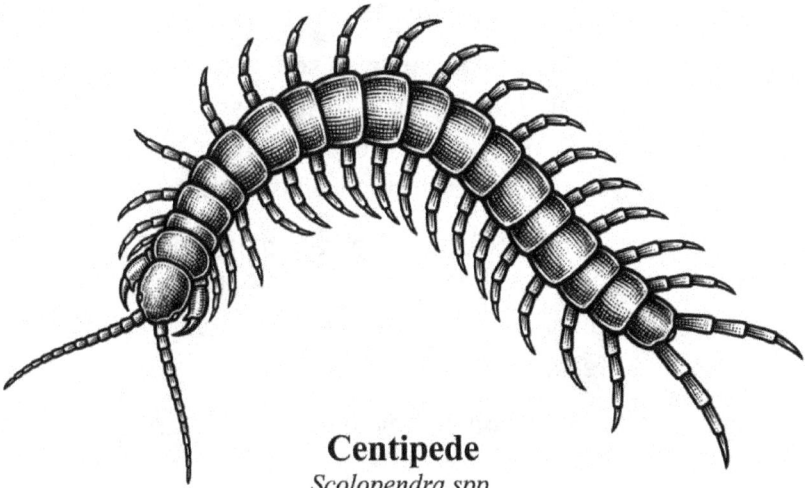

Centipede
Scolopendra spp.

COMMON LOCATION: Worldwide
Uses venomous front legs and flexible bodies to hunt in narrow spaces.

FUN & WEIRD FACTS

MIND-BLOWN MOMENT

The trap-jaw ant Strumigenys (a tiny rainforest ant with spring-loaded jaws) snaps its mandibles shut so fast that it can fling itself backward. It uses this sudden jump to escape predators that tower over it. This move is so quick that high-speed cameras struggle to capture it.

The assassin bug nymph (a young predator with a straw-like piercing mouth) waits under leaves until another insect walks near. It injects venom that turns the prey's insides into liquid. The nymph then drinks the liquefied tissues.

The Brazilian ambush beetle larva (a soft-bodied hunter hidden under soil in the Amazon) builds funnel-shaped traps in loose dirt. When an insect slides in, the larva lunges upward and grabs it. It drags the prey underground to feed safely.

The velvet worm (a rainforest invertebrate that shoots glue from its face) fires two sticky streams that cross in midair. The slime hardens instantly and traps the prey. The velvet worm then moves in slowly to feed.

The jungle pseudoscorpion (a tiny arachnid with strong pincers but no tail stinger) rides on larger insects to reach new hiding places. This helps it find fresh food inside logs and leaf piles. It hunts mites and springtails with surprising strength.

The parasitic phorid fly (a small fly that targets ants) injects its egg behind an ant's head. The larva grows inside and slowly weakens the ant until the head falls off. The larva finishes feeding inside the empty head.

The spiny leaf-mimic katydid nymph (a young katydid shaped like a ragged leaf piece) blends perfectly into jungle foliage. Its uneven edges match damaged leaves so well that predators often miss it. This disguise works until the nymph grows too large.

The trapdoor spiderling (a tiny spider that builds soil-covered burrows) creates a miniature silk-and-dirt door. It hides beneath the lid and waits for vibrations. When prey walks close, it snaps the door open and attacks.

The tropical rove beetle (a fast beetle that hunts in leaf litter) releases a chemical that weakens millipedes. The millipede stiffens and cannot defend itself. The beetle then feeds safely.

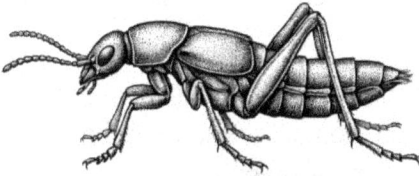

Tropical Rove Beetle
Staphylinidae

COMMON LOCATION: Tropical forests worldwide.

DESCRIPTION: Tropical rove beetles hunt rapidly through leaf litter and use chemical secretions to weaken millipedes. The toxin causes the millipede to stiffen and lose its defensive curl, allowing the beetle to feed on prey that would normally be too well armored to attack.

The big-headed ant major (a worker ant with a huge shield-like head) blocks narrow nest entrances with its head. Only ants from the same colony may pass by tapping gently. This protects hidden tunnels from intruders.

The Amazonian termite-raider centipede (a long predator that hunts termites underground) senses tiny vibrations through soil. It bites through termite walls to reach the colony. Its venom spreads quickly and disables many termites at once.

The moss mantis (a small mantis that looks like a patch of moss) stays still on tree bark. Its body texture blends into the damp forest surface. Insects often come close without noticing it.

The parasitoid wasp Glyptapanteles (a wasp whose larvae grow inside caterpillars) lays many eggs inside one host. After emerging, the larvae spin cocoons while the weakened caterpillar guards them. The caterpillar dies shortly afterward.

The Panamanian ant-snare beetle larva (a larva that builds sticky silk nets under roots) stretches thin threads across soil gaps. Ants running past get stuck in the silk. The larva senses the vibrations and pulls them underground.

The devil's garden ant (an ant that clears plants using formic acid) poisons unwanted vegetation around its nests. This creates open patches in the forest called devil's gardens.

The woodlouse spider juvenile (a young spider that hunts isopods) has long, curved fangs that pierce tough shells. It delivers a fast bite that stops its prey. These spiders grow more specialized with age.

The thorn mimic treehopper nymph (a young insect shaped like a plant thorn) sits motionless on stems. Its shape and color fool predators completely. Only movement reveals it is alive.

The contortion roach (a small roach with flexible armor plates) bends sharply to slip into cracks thinner than a fingernail. This helps it escape spiders and centipedes. It relies on tight spaces for survival.

Contortion Cockroach
Periplaneta spp.

COMMON LOCATION: Tropical and subtropical regions worldwide.

DESCRIPTION: Some cockroaches have flexible armor plates that allow them to bend and flatten their bodies, slipping into cracks thinner than a fingernail. This extreme flexibility helps them escape predators like spiders and centipedes by vanishing into tight spaces others cannot enter.

The vampire moth Calyptra (a moth capable of piercing skin) usually feeds on fruit. But it can also drink mammal blood using tiny hooks on its mouthparts. This skill is rare among moths.

The jaguar ant (a solitary predatory ant known for silent movement) hunts alone through leaf litter. It disables insects larger than itself with quick bites. Its hunting style is quiet and precise.

The leaf-litter centipede Cryptops (a blind centipede that hunts using vibrations) senses even tiny ground movements. It uses touch instead of sight to find hidden insects. Its reactions are extremely fast.

The termite trap fungus (a fungus that imitates termite scents) lures termites into hidden chambers. Once inside, they stick to the fungal surface. The fungus slowly absorbs their nutrients.

The stilt-legged fly nymph (a young fly with long, thin sensing legs) stands above leaf litter and detects air vibrations. These vibrations reveal insects moving below. It drops down quickly to strike.

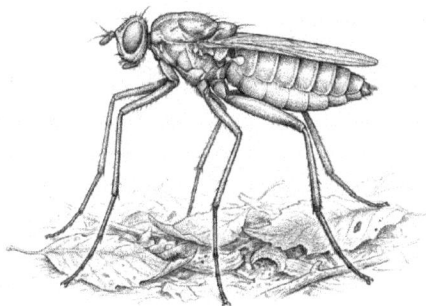

Stilt-Legged Fly Nymph
Micropezidae (nymph stage)

COMMON LOCATION:
Forests and woodland leaf litter worldwide.

DESCRIPTION:
The stilt-legged fly nymph stands high above the ground on long, thin sensing legs that detect tiny air vibrations. Movements below the leaf litter reveal hidden insects. In a sudden drop, the nymph strikes downward to capture prey with speed and precision.

The minnow beetle larva (an aquatic larva found in jungle streams) grabs insects and spins them rapidly. This spinning tears the prey's soft body apart. It then feeds on the released nutrients.

The leaf-cutter ant microcleaner (a tiny ant that rides on larger ants) removes harmful spores from workers. This prevents dangerous fungi from spreading. The entire colony depends on these cleaners.

The spined orb-weaver spiderling (a tiny spider with defensive spikes) builds miniature shiny webs. Because the webs are small, insects often run straight into them. The spiderling uses these catches for early hunting practice.

The army ant queen's retinue larvae (young ants that cling to workers for transport) ride on foraging ants to reach fresh kills. This keeps them safe from ground predators. They feed on whatever prey the workers bring back.

The assassin lacewing larva (a small predator that covers itself with debris and old prey parts) moves across the forest floor disguised as dirt. Insects walk near without realizing it is alive. The larva snaps forward and holds tight until the prey stops moving.

The micro-tarantula juvenile (a very small rainforest tarantula) digs tunnels that collapse behind it. This prevents predators from following. Only when it resurfaces does it rebuild an entrance.

The spitting spider nymph (a young spider that spits sticky silk mixed with venom) fires a zigzag spray that spreads as it travels. The silk tightens around prey like a net. The venom works almost instantly.

The stingless bee soldier (a large-jawed worker bee) guards narrow entrance tunnels in tree nests. When intruders appear, soldiers rush forward and bite repeatedly. Their teamwork makes the colony very hard to invade.

Stingless Bee Soldier
Meliponini

COMMON LOCATION:
Tropical forests worldwide.

DESCRIPTION:
Some stingless bees produce large-jawed worker soldiers that guard narrow entrance tunnels in tree nests. When intruders approach, these defenders rush forward and bite repeatedly, using teamwork rather than stings to protect the colony.

The jungle click mite (a tiny mite with a spring-like plate) can launch itself several inches into the air. This huge jump helps it escape predators quickly. It resets the plate for another jump within seconds.

The false-tiger roach (a roach that mimics aggressive beetles) uses bold stripes and posture to scare predators. Most animals avoid it because they think it is dangerous. A good disguise.

The gall-forming psyllid (a tiny sap-feeding insect that makes leaves form round shelters) injects chemicals into leaves. The leaf swells into a hollow ball that protects the psyllid. These galls shield it from nearly all predators.

The mineral-scraper beetle (a beetle with zinc-strengthened jaws) cuts through hard soil crusts to reach fungi. Its reinforced mouthparts act like tiny metal blades. This lets it feed in places other insects cannot reach.

The feather-legged assassin bug nymph (a predator with fluffy leg extensions) waves its soft legs to attract ants. When an ant approaches, the nymph delivers a fast toxin. It feeds safely once the prey is still.

The soil centipede Geophilus (a thin centipede that lives under leaf litter) releases a peppery smell when threatened. This odor warns off spiders and beetles. The centipede quickly disappears into the soil.

The jungle earwig (a small insect with strong pincers) guards its eggs inside underground chambers. It fights off anything that comes near. The young stay with their mother until they molt.

The moss-ghost springtail (a tiny springtail with a transparent body) becomes nearly invisible under filtered sunlight. Predators easily overlook it. This makes it one of the safest insects in the leaf litter.

The parasitic 'strepsipteran' (a tiny insect that lives inside other insects) forces leafhoppers to perch in exposed places. When the parasite emerges, the host dies. It then searches for new insects to infect.

The 'fer-de-lance' flea (a flea often found near snake resting spots) feeds on animals hiding in leaf litter. It can track both warm and cold-blooded hosts. This makes it unusually adaptable.

The jungle antlion microlarva (a larva that builds extremely small sand traps) digs pits so tiny they fit on a fingertip. Ants and mites slide into the loose grains and cannot escape. The larva waits at the bottom to bite passing prey.

The fungus-grooming roach (a roach that removes spores from other insects) picks fungal spores off beetles and ants. It also eats weakened insects it finds. This helps keep the forest floor balanced.

The miniature whip-scorpion (a rice-sized arachnid with strong claws) captures insects by grabbing and crushing them. Despite its size, it has surprising strength. It hides under bark during the day.

Miniature Whip Scorpion
Schizomida (short-tailed whip scorpions)

COMMON LOCATION: Tropical and subtropical forests worldwide.

DESCRIPTION: These rice-sized arachnids use powerful claw-like pedipalps to grab and crush small insects. Despite their tiny size, miniature whip scorpions have surprising strength and spend daylight hours hidden beneath bark and leaf litter.

The leaf-mimic stick insect nymph (a small insect that copies leaf colors and shapes) eats moss that stains its body green. This natural dye helps it blend into the jungle floor. Its camouflage improves as it grows.

The amber mound termite (a termite that builds hard, mineral-rich mounds) mixes soil with saliva to form walls as tough as stone. These mounds protect against predators and flooding. They can last for many years.

The bone-house wasp larva (a larva raised in a nest lined with dead ants) grows inside a protective chamber. The dead ants hide its scent from predators. The adult wasp emerges safely from this unusual nursery.

MYTHS - BUSTED

Some people believe tiny jungle predators hunt at night because they can see in total darkness, but their true mastery lies in sensing the unseen. **Micro-predators** (microscopic hunters in tropical leaf litter) track prey through vibrations, scent molecules, and even the air pressure from passing footsteps. In a world too cluttered for sight, they are guided by touch so precise they could strike prey blind-and still never miss.

Roaches in the rainforest survive not because their armor is tough, but because their bodies are built like flexible origami. **Rainforest roaches** (small nocturnal scavenger beetles) flatten and fold until they can vanish into a crack thinner than a fingernail. Their secret power is bendability, letting them escape jaws, floods, and falling debris that would crush other insects.

Tiny parasitic flies are rumored to command entire ant colonies, but each one conquers only a single victim. **Phorid flies** (parasitic flies that infect ants) lay eggs inside one ant, and when the larva grows, it rewires the ant's behavior so completely that it wanders away alone. The sight of multiple "zombie" ants has tricked many into thinking the whole colony is possessed at once.

Miniature mantises hide in moss not because they're weak, but because they are ambush artists. **Moss mantises** (tiny camouflaged mantises from tropical forests) disguise themselves among green filaments until an unsuspecting insect passes their reach. One lightning strike later, the killer moss comes alive, and something else disappears.

Young jungle spiders spin webs that look like fine lace, leading people to think they're learning. In truth, these **spiderlings** (juvenile spiders from rainforest canopies) design their webs to match both their size and the delicate prey they stalk. Their traps are miniature masterpieces-scaled down, not dumbed down.

Jungle earwigs have a reputation for pinching people, but they're actually gentle guardians of their nests. **Earwigs** (tropical insects with curved pincers at the abdomen) use those claws to defend eggs, tidy burrows, and challenge intruders only when cornered. If they pinch a finger, it's self-defense, not malice-a tiny monster just protecting its home.

"Hmmmm, this date might not go so good"

© MIND-BLOWN^{tm} Cartoons

Bug Smiles

- What is a mosquito's worst fear? – The S.W.A.T Team.

- How do you measure a mosquito's harddrive? – bug bytes.

- What do you get if you cross the Lone Ranger with an insect? – The Masked-quito!

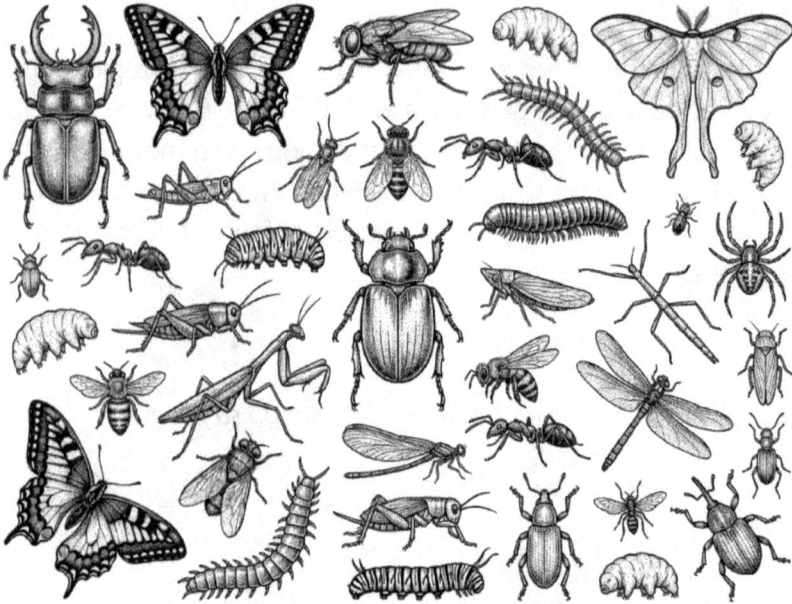

🔍 MICRO-JUNGLE MONSTERS

Step into a jungle so small most humans never notice it. Here, a grain of sand becomes a boulder, a single leaf turns into a skyscraper, and survival happens at lightning speed.

Micro-Jungle Tiny Monsters stalk, ambush, leap, and disguise themselves in a world where every shadow hides danger and every movement matters.

These creatures live out full-scale dramas beneath our feet — fierce battles, clever escapes, and impossible athletic feats — all unfolding in places we walk past every day without a second glance.

LEGENDS

LEGENDS

The Devil's Garden Ants of the Western Amazon

Travelers along the Rio Napo in Peru tell of clearings called "devil's gardens," where only one plant grows and the rest of the forest seems afraid to enter. Locals say the lemon ants living there protect these patches with unusual aggression, chasing off anything that disturbs their chosen trees. Scientists later confirmed that the ants actually poison the surrounding plants, but the eerie emptiness of these gardens still feels supernatural to many who pass through them. Some guides swear you can hear the ants before you see them.

The Firefly Funeral Lights of the Great Smoky Mountains

In Appalachian folklore, synchronized *Photinus carolinus* fireflies were once said to gather only where someone had recently died, forming bright, pulsing "soul lights" over the ground. Modern researchers know their flashing is a mating display, but older residents insist the insects appear early in places touched by grief. Hikers claim the forest grows unusually still when the lights begin to pulse in perfect unison, and many admit the timing has felt unsettling more than once.

The Death-Watch Beetles of Yorkshire, England

For centuries, families in old English farmhouses believed the ticking sound of the death-watch beetle signaled coming misfortune. The beetle makes its knocking sounds by tapping its head on wood, but generations interpreted it as a supernatural warning. During wartime blackouts, people reported the tapping grew louder, as if the beetles sensed the fear in the home. Restoration workers today still say the sound feels strangely deliberate in empty rooms.

The Weeping Cicadas of Northern Thailand

Villagers near Chiang Mai tell of a forest hillside where rain cicadas begin calling long before clouds appear. According to local belief, the cicadas "weep" for travelers who ignore nature's warnings and continue up the mountain. Hikers report the insects fall silent the moment the first raindrops hit the leaves, adding to the eerie sense that the calls were a deliberate alert.

The Glowworm Path of New Zealand's Waitomo Forest

Visitors trekking near the Waitomo cave systems describe nights when hundreds of *Arachnocampa* glowworms appear outside the caves, forming a faint trail of blue light through the forest. Māori legends say the lights guide lost ancestors home, while modern hikers claim the glowworms appear only when someone has strayed off the main path. Scientists believe humidity draws them out, but the perfect, path-like shapes remain unexplained.

The Crying Beetles of the Pacific Northwest

Residents near old logging towns in Washington State whisper about soft, tapping cries heard inside abandoned cabins. The sound comes from anobiid beetles striking their jaws against wood, though locals long believed it signaled a troubled spirit. Hikers claim the tapping grows louder before storms, echoing through rafters like someone knocking for help. Even forestry workers admit the sound feels strangely mournful in empty buildings.

The Warning Ants of Borneo's Rainforest Floor

Indigenous guides in Sabah tell stories of giant forest ants, *Camponotus gigas*, that act as silent guardians of certain trails. The ants appear in large numbers when a predator such as a clouded leopard is near, forming restless lines that block the path. Scientists credit chemical cues, but the timing is so precise that even researchers admit it feels intentional.

DID YOU KNOW?

Did you know? Some **army ants** create living bridges using their own bodies, linking themselves together so entire colonies can cross gaps in the rainforest floor. The ants constantly adjust the bridge's shape and thickness according to traffic, forming a structure that flows and flexes in real time. Scientists study these bridges as examples of collective engineering-tens of thousands building something no single **ant** could ever plan alone.

Did you know? The **trap-jaw ant** doesn't just catch prey with its lightning-fast jaws-it can fire itself backward to escape danger. The explosive snap launches the **ant** through the air faster than many larger animals can move. Even high-speed cameras struggle to freeze the moment it takes flight.

Did you know? Certain **rainforest millipedes** store cyanide-based chemicals in their bodies for defense, releasing a bitter almond scent when threatened. These toxins discourage predators from foraging where millipedes are dense, shaping the ecosystem from the ground up. A few grams of tiny **millipedes** can influence where much larger hunters choose to live.

Did you know? Some **leafcutter ants** maintain vast underground fungus farms that can spread wider than a small bedroom. They feed the gardens with fresh leaves and remove harmful molds to keep the crop healthy. Biologists compare the **ants'** system to early human agriculture-ancient farming on a miniature scale.

🐛 **Did you know?** The **larvae of lacewing insects** disguise themselves with dust, leaf bits, and even the dried remains of their prey. This "trash armor" lets them ambush victims unnoticed and protects them from **ant** defenses. To predators, a **lacewing larva** just looks like a harmless speck of forest debris.

🐛 **Did you know?** Many **orb-weaver spiders** recycle their webs by eating the silk each night before spinning a new one at dawn. Reused proteins create stronger, cleaner strands that stand up to wind and rain. It's nature's ultimate zero-waste design.

🐛 **Did you know?** Tiny **pseudoscorpions** hitch rides on beetles and flies to explore new logs and leaf piles. This harmless method, called phoresy, turns larger insects into living transportation. A single **pseudoscorpion** may travel miles while never taking a step.

🐛 **Did you know?**
Some **bioluminescent beetle larvae** glow continuously under tropical leaf litter. Their faint green light warns predators that they taste bad and also helps them find safe, open hunting spots. The forest floor becomes a dim map of predator and prey written in light.

🐛 **Did you know?** Many **house spiders** sense prey not only through web vibrations but also through subtle changes in air pressure. This ability lets them anticipate an insect's approach before a single thread stirs. They prefer corners where air currents converge-perfect ambush zones.

🐛 **Did you know?** Common **cockroaches** feel danger before they see it. Tiny sensory hairs on their legs respond instantly to moving air, alerting them to footsteps or shifting shadows from several feet away. That's why **roaches** vanish the moment a light flicks on.

COMIC 1 — ARMY ANT BRIDGE

Hold steady. Traffic's heavy today.

I signed up to be an ant... not infrastructure.

COMIC 2 — TRAP JAW ANT ESCAPE

So... you just bite things really fast?

Yes.

He also flies.

MIND-BLOWN™ Cartoons

COMIC 1 — PSEUDOSCORPION RIDESHARE

"Are you getting off at some point?"

"I don't actually know. I just go where you go."

"It's quiet. I think we're safe."

COMIC 2 — COCKROACHES & AIR

"RUN."

MIND-BLOWN™ Cartoons

STORY MOMENT

STORY MOMENT

The Floor That Moved

You almost missed it at first-the forest floor here always seems to breathe. Beneath the green canopy, everything crawls, shifts, and exhales in its own rhythm. But tonight the rhythm felt different, a quiet that didn't match the usual chaos of crickets and dripping vines. When you knelt beside a moss-covered log, the ground beneath your knee gave an unexpected sigh, a subtle rise and fall, as if something just below the surface had taken a deep, deliberate breath.

The litter moved gently at first, no louder than a whisper of sand, then a slow wave rippled outward, setting tiny seeds shaking on their stems. You placed a hand to the soil and felt warmth pass through the thin layer of leaves-a pulse, soft and timed, as if the earth itself had a heartbeat. Out of the corner of your eye, a small glint appeared: a **beetle**, bronze-green and polished, pushing through the debris with astonishing precision. Another followed. Then another.

Soon dozens of **beetles** surfaced at once, climbing from the dark in perfect rhythm. They didn't scatter or flee the light from your lamp. Instead, they pressed together in a widening circle, each moving with identical purpose. Antennae rose and fell as if responding to an invisible conductor. The ring tightened, the soil in the center trembling slightly, and a faint hum began-a vibration that climbed up through your fingertips, through your wrist, through your spine until it found your heart and matched its beat.

The air grew thick and wet with the scent of metal, sap, and distant thunder. Leaves rustled without wind. The jungle seemed to pause, holding its collective breath. Beneath your palm, the central patch of dirt bulged-once, twice-as if a lung pushed from below. Even the smallest roots trembled. For one disorienting instant you saw the entire forest connected: trees, moss, and insects pulsing together in one shared signal, the ground alive and aware.

The hum deepened to a tone almost too low for hearing, strong enough that the edges of your vision trembled.
The **beetles** leaned inward, wings half-unfolded, as though awaiting a final cue. Then silence cracked the moment in two. A single vibration fluttered across the soil-and everything stopped. The circle dissolved, insects slipping back beneath the surface one by one, their passage leaving faint tunnels that collapsed behind them.

The heat dropped away. The metallic scent faded to simple rain-soaked earth. You waited, barely breathing, the beam of your flashlight catching motes of mist where the creatures had gathered. Nothing moved now. Yet the impression lingered-that the jungle floor had turned itself inside out for you, revealing a secret rhythm older than speech. When you finally rose and brushed the dirt from your hands, the forest had already resumed its steady breathing, pretending innocence, as if it hadn't just shown you how perfectly the smallest monsters can move as one.

FUN QUIZ

1. What adaptation allows the trap-jaw ant to launch itself backward to escape predators?

Explain how this movement works.

2. True or False: Pseudoscorpions spread through the forest mostly by walking long distances.

Write "true" or "false."

3. Which behavior makes lacewing larvae difficult for ants to detect?

A. Their bright colors

B. Their debris camouflage

C. Their speed

D. Their ability to jump

4. What chemical defense do certain rainforest millipedes use, and how does it influence where predators hunt?

5. True or False: Orb-weaver spiders build a new web every night and recycle the old one.

Write "true" or "false."

6. Why do bioluminescent beetle larvae glow beneath the leaf litter?

Explain at least one advantage of this behavior

QUIZ ANSWERS

1. The trap-jaw ant snaps its mandibles shut with explosive speed, generating enough force to launch itself backward as an escape. Its jaws act like a spring-loaded mechanism.

2. False. Pseudoscorpions spread by hitching rides on larger insects.

3. B - Their debris camouflage.

4. They release cyanide-based chemicals that give off an almond-like smell; predators avoid these areas, shaping where they hunt.

5. True. Orb-weaver spiders recycle their silk to build a fresh web each night.

6. The glow warns predators they may be toxic or unappealing; it also helps them find safe hunting spaces in the dark.

CHAPTER 8

8- Bugs that Break the Rules of Life

Extreme survival, freezing, cloning, longevity

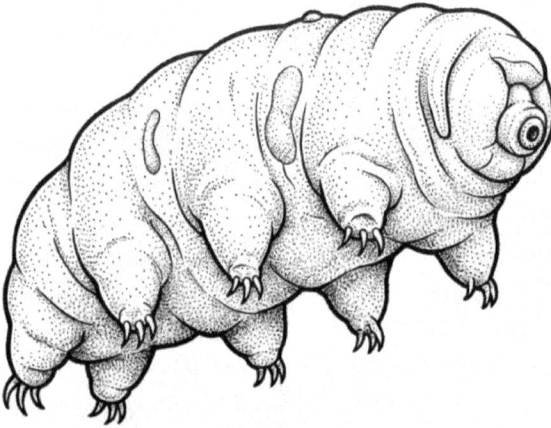

Tardigrade (Water Bear)
Tardigrada phylum

COMMON LOCATION: Worldwide
Can survive extreme heat, cold, pressure, radiation, and even space.

FUN & WEIRD FACTS

MIND-BLOWN MOMENT

The emerald cockroach wasp *(a parasitic wasp from Southeas t Asia)* can disable one precise part of a cockroach's brain-the area that triggers escape. The victim stays alive and able to walk but loses all will to run. The **wasp** then leads it calmly into a burrow for its larva. It's biological mind-control perfected.

The American cockroach *(common worldwide)* can hold its breath for over thirty minutes. By closing spiracles to retain moisture and block toxins, the **cockroach** becomes almost fumigation-proof. Its respiratory control surpasses expectations for something so small.

The Antarctic springtail *(soil creature from Antarctic islands)* survives brutal cold that would freeze nearly all land animals. Natural antifreeze in its body prevents ice from forming within its cells. Even under solid snow, the **springtail** keeps moving-an insect unlocked from winter's rules.

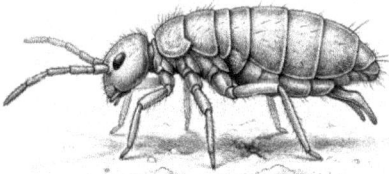

Antarctic Springtail

Collembola (cold-adapted soil species)

COMMON LOCATION:
Antarctic coastal islands and polar soil beneath snow and ice.

DESCRIPTION:
The Antarctic springtail survives temperatures that would instantly freeze nearly all land animals. Natural antifreeze chemicals in its body prevent ice crystals from forming inside its cells. Even sealed beneath snow and ice, the springtail remains active—an insect operating beyond winter's normal limits.

The house spider *(a widespread indoor species)* can rebuild a symmetrical web after losing a leg. It re-calculates line angles and silk spacing on the fly. The **spider's** structural problem-solving rivals advanced robotics repair.

The bombardier beetle *(a ground beetle found in forests world wide)* fires boiling liquid from its abdomen using chemistry. Two benign compounds mix in a reinforced chamber to erupt in pops of steam and heat. This **beetle's** weapon functions like a microscopic cannon.

The woolly aphid *(a sap-feeding insect covered in cotton-like w ax)* protects itself with sticky threads that jam predators' jaws. The wax strands snare **ants** and **ladybugs**, letting the aphid retreat. Its fluff isn't decoration-it's armor.

The trap-jaw ant *(found in tropical forests)* snaps its jaws shut with record-setting velocity. The recoil launches the **ant** into mid-air to evade danger faster than an eye blink. Few creatures break physical limits like this jawed jumper.

The housefly *(common worldwide)* processes vision about seven times faster than humans. A swatting hand looks to a **fly** as if moving through syrup, making escape effortless. Its fast-time perception turns seconds into survival.

The velvet ant *(a wingless wasp with bright red coloring)* wears an exoskeleton so tough that even hammers barely dent it. Its sting is one of the most painful in nature. Despite its name, the **velvet ant** is a wasp built like tank and torch combined.

The Argentine ant *(an invasive species now found worldwide)* ignores territorial boundaries, merging into continent-wide supercolonies. The **ants** share chemical identity, so members from distant nests cooperate instead of fight. Their social reboot rewrites the rules of colony warfare.

The New Zealand glowworm *(a cave-dwelling larva)* casts sticky silk threads that shine like a night sky of green stars. Flying insects climb toward the light and become hanging meals. The **glowworm** hunts with a galaxy.

The house centipede *(a fast-moving indoor predator)* hunts prey much larger than itself. Its lashing legs and venomous grip take down **roaches**, **spiders**, and **silverfish** with ease. When one **centipede** moves in, household ecosystems reorder overnight.

The woolly bear caterpillar *(found in cold northern climates)* freezes solid each winter, its cells protected by cryoprotectants. In spring it thaws and crawls away alive. The **caterpillar** is the insect world's re-animation artist.

The Japanese giant hornet *(native to East Asia)* warms flight muscles before takeoff, becoming the first predator airborne on chilly mornings. This pre-heating power gives the **hornet** a lethal head start while others still shiver.

Japanese Giant Hornet
Vespa mandarinia (adult)

COMMON LOCATION:
Wooded and rural areas of East Asia, especially Japan.

DESCRIPTION:
The Japanese giant hornet warms its flight muscles before takeoff by vibrating its body to generate heat. This allows it to fly during cool mornings when most insects remain inactive. The hornet gains an early hunting advantage by becoming airborne while others are still slowed by the cold.

The whirligig beetle *(a surface swimmer of pond water)* sees above and below at once through divided eyes. The **beetle** scans air predators and underwater hunters simultaneously while spinning in dizzying circles. Evolution fitted it with natural periscopes.

The assassin bug *(a stealthy predator found worldwide)* injects enzymes that turn prey to liquid and drinks it. It eats without chewing, melting its victim first. No dining etiquette equals that of the **assassin bug.**

The jumping bristletail *(an ancient insect relative from rocky la nds)* flings itself several inches into the air using abdominal muscles. By catapulting backward, the **bristletail** escapes danger despite having no wings at all-prehistoric escape tech still working.

The dragonfly nymph *(a juvenile dragonfly living underwater)* uses jet propulsion to lunge. It expels water forcefully from its rear, shooting after prey or away from predators. This hydraulic burst makes the **nymph** a torpedo of the pond.

The flea *(a parasite on pets and wildlife)* leaps many times its body length thanks to the elastic protein resilin. Its spring-loaded legs let the **flea** accelerate faster than a rocket at launch scale. Physics bends to its tiny knees.

The Mexican jumping bean larva *(a moth larva inside seeds)* rocks its wooden pod by shifting rhythmically inside. The moving seed "jumps" to stay cool and out of reach of birds. Together, the **larva** and the seed create a living trick.

The death-watch beetle *(a wood-boring insect from old buildin gs)* taps its head against wood to signal mates. The sound once terrified households as a death omen during midnight quiet. The **beetle's** love knock became a legend.

The tarantula hawk wasp *(a large wasp from the American So uthwest)* paralyzes tarantulas with a single sting and drags them underground. The **wasp** hunts giants, proving brain and venom outweigh scale.

The snow fly *(a wingless winter fly from northern forests)* walks on snow where others freeze. Antifreeze-like compounds inside the **fly** let it live entire seasons on ice, breaking the insect rulebook for temperature.

The leafcutter ant *(common in Central and South America)* cultivates underground fungus farms fed by chewed leaf bits. Their agriculture predates humans by millions of years. The **ants** are miniature farmers running ancient greenhouses.

The house dust mite *(a microscopic creature in bedding and car pets)* lives entirely on human skin flakes and drinks water from the air. Invisible armies of **mites** thrive in places we think are dry. Few organisms exploit our habits so completely.

HOUSE DUST MITE
Dermatophagoides (adult)

COMMON LOCATION:
Human dwellings worldwide, especially bedding, mattresses, and carpets.

DESCRIPTION:
The house dust mite lives entirely on flakes of human skin and absorbs water directly from the air. Even spaces that feel dry contain enough microscopic moisture to sustain large populations. Few organism: are so completely adapted to human living environments while remaining almost entirely unseen.

The silverfish *(a primitive household insect)* continues to molt even after adulthood. The **silverfish** never stops growing, shedding skins through old age-an ancient survival tactic granting unexpected longevity.

The sharpshooter *(a leafhopper from warm regions)* fires waste droplets at great speed using a microscopic catapult. This keeps it clean while feeding constantly on sap. The **sharpshooter's** plumbing impressed engineers studying efficiency.

The camel cricket *(a long-legged basement insect)* leaps several feet with powerful hind legs, startling predators-and people. Despite its alien look, the **cricket** is harmless and oddly graceful underground.

The spittlebug *(a plant-feeding insect common in gardens)* hides in frothy "nests" that act as insulation and camouflage. Instead of armor plates, the **spittlebug** relies on bubbles-a soft fortress against heat and enemies.

The glow-spot roach *(a bioluminescent roach from South Amer ica)* displays two green lights on its thorax that imitate toxic beetles. Predators mistake the **roach** for something dangerous and move on. It glows to survive.

The water boatman *(a small aquatic insect)* creates underwater songs louder than many birds by rubbing its body parts together. The sound carries across ponds, showing that the **boatman** breaks every rule about small size meaning quiet voice.

The antlion larva *(a sand-dwelling predator)* builds perfect funnel pits that collapse under stray footsteps. Any **ant** or beetle sliding in becomes lunch. Its obsessive geometry is instinctive genius.

The rove beetle *(a slender beetle found worldwide)* can sacrifice the tip of its abdomen to distract attackers. The twitching piece confuses predators while the **beetle** escapes-then it simply regrows what it lost.

The carpenter ant *(living in forests and homes)* communicates through vibrations sent along wood. Alarms travel through beams faster than sound, giving the **colony** instant warnings through its wooden network.

The parasitoid tachinid fly *(a fly family found worldwide)* lays eggs on caterpillars that the larvae slowly consume from inside. They avoid vital organs until maturity, keeping the host alive. Evolution rarely designs cruelty this efficient.

The Hawaiian happy-face spider *(a tiny rainforest species)* wears cheerful, shifting smile-like patterns that confuse hunters. The **spider** can even alter its markings slightly-a morphing mask for survival.

Hawaiian Happy Face Spider
Theridion grallator (adult)

COMMON LOCATION:
Native to the rainforests of the Hawaiian Islands.

DESCRIPTION:
The Hawaiian happy face spider displays bright, face-like patterns on its abdomen that vary between individuals. These markings disrupt visual recognition and confuse predators that hunt by sight. Some individuals can subtly alter pattern intensity over time, creating a shifting disguise that improves survival.

The soldier termite *(from tropical regions)* is born with a colossal head shaped like a plug solely to block tunnels. During invasion, the **termite** uses itself as a living door. Few body plans are more purpose-built.

The water strider *(a pond surface insect)* walks effortlessly on water using hydrophobic leg hairs that trap air. The **strider** skims without breaking the surface film, turning physics into playground.

The firebrat *(a heat-loving household insect)* thrives in attics, ovens, and boiler rooms near 100 °F. Proteins in the **firebrat** resist heat damage, letting it live where most insects would shrivel.

The owl fly larva *(a predator from warm regions)* buries itself in sand with jaws poised upward. When prey passes, the **larva** explodes upward like living shrapnel. Its ambush rivals the antlion's perfection.

The fruit fly *(Drosophila melanogaster* –* a model research species) * evolves visibly within weeks, adapting across generations before our eyes. Scientists use **fruit flies** to watch evolution in real time-a laboratory of life compressed.

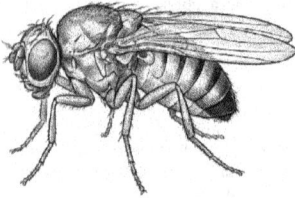

Fruit Fly

Drosophila melanogaster (adult)

COMMON LOCATION:
Worldwide in human-associated human-associated environments; laboratories globally as a research model species.

DESCRIPTION: The fruit fly evolves rapidly, showing visible genetic and physical changes within just a few generations. With a short life cycle, entire evolutionary shifts can be be observed in weeks rather than centuries. Scientists use fruit flies to study inheritance, adaptation, and evolution in real time—a compressed laboratory of life.

The scale insect *(a tree-dwelling plant parasite)* secretes a hardened wax shield thicker than its own body. Beneath the armor, the **insect** drinks sap in safety for weeks. Its private fortress beats most armor-clad animals.

The backswimmer *(a small aquatic predator)* swims belly-up using air-packed legs for buoyancy. The **backswimmer** hunts upside down, watching prey silhouetted against the sky-a reversed world under glass.

The lacewing larva *(a small predatory insect)* stacks the remains of its victims on its back for camouflage. To **ants**, it looks like a random dirt clump, not a killer. It hides in plain sight under trophies.

MYTHS - BUSTED

MYTH: Insects can't survive without oxygen. That sounds logical-until you meet bugs that shut down breathing entirely and just wait it out. Some insects survive hours, days, or even months with no oxygen at all by switching into a near-paused metabolic state that would kill most animals outright.

MYTH: Nothing alive can survive being frozen solid. Certain insects laugh at this rule by turning their own bodies into biological antifreeze. Species like freeze-tolerant beetles and moth larvae can let their cells freeze completely, then thaw back to life as if nothing happened.

MYTH: Losing your head means instant death. Not for insects. Some can function for days or even weeks after decapitation because their vital systems aren't controlled by a single brain-and they don't need constant oxygen delivery to survive.

MYTH: You must eat regularly or you'll die. Some insects can go for shockingly long stretches without food, surviving purely on stored energy and ultra-efficient metabolism. A few species can last months-and in extreme cases over a year-without taking a single bite.

MYTH: Life can't survive in extreme radiation. Insects like cockroaches and certain fly larvae can withstand radiation doses that would be fatal to humans many times over. Their cells repair DNA damage far more effectively, breaking one of biology's most feared rules.

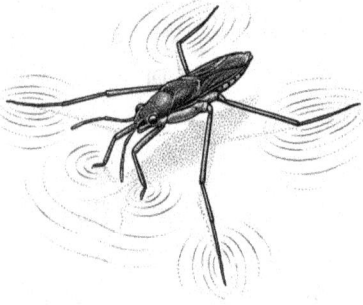

Water Strider
Gerridae family
Walks on water using surface tension.

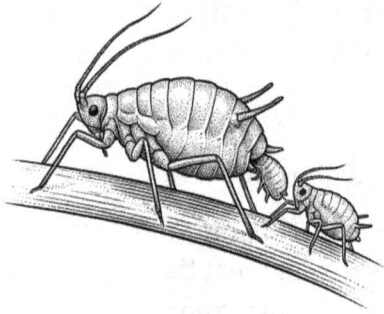

Aphid
Aphidoidea
Females give birth to live clones already pregnant.

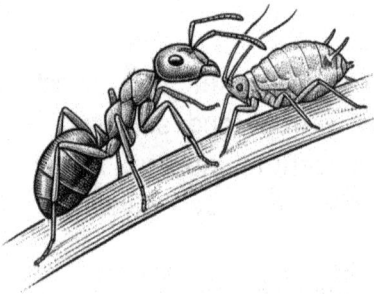

Ant Farming Aphids
Formicidae
Ants farm aphids like livestock.

Metamorphosis
Lepidoptera
Body dissolves and rebuilds during transformation.

LEGENDS

LEGENDS

The Beetles That Crawled from the Pharaohs

Ancient Egyptian legends tell of beetle-like creatures emerging from sealed tombs long after burial chambers were closed (ancient Egypt, North Africa). Locals believed these insects were guardians of the dead, awakened when sacred rules were broken, and some early tomb workers claimed swarms appeared where no insects should have survived.

The Fire Ants of the Devil's Fields

In parts of South America, stories describe vast ant colonies living in scorched, lifeless ground where nothing else grows (Amazon Basin, South America). According to legend, these ants were created by spirits to punish the land, surviving heat and toxins that kill all other creatures.

The Whispering Locust Clouds

Medieval travelers wrote of locust swarms that didn't just devour crops but seemed to make a low, constant whispering sound as they passed overhead (Middle East and North Africa). Folklore claimed the insects carried voices of the dead, warning villages before famine struck.

The Ice Worms of the Frozen Peaks

Mountain legends describe strange worm-like insects that crawl across glaciers at night and vanish by sunrise (Alaska and Himalayan regions). People believed these creatures were born from ice itself, feeding on cold and dying instantly if touched by warmth.

The Undying Roaches of the Atomic Zone
After nuclear testing, rumors spread of cockroaches thriving inside highly radioactive areas where humans couldn't survive (Nevada desert, USA). Locals claimed the insects were altered by the blast, becoming symbols of life that refuses to obey nature's limits.

The Beetle That Walks Through Fire
Stories from desert regions tell of a shiny black beetle seen calmly crossing burning brush and lava-hot ground (Australian and African deserts). Witnesses believed the insect was protected by fire spirits, immune to heat meant to destroy all living things.

The Moth That Eats Ghost Light
In parts of Europe, folklore describes pale moths drawn not to lamps, but to glowing orbs seen in graveyards and ruins (Western and Eastern Europe). Legends say these moths feed on restless energy and vanish when the spirits finally move on.

The Ants That Built the Lost City
Ancient legends claim entire stone ruins were slowly assembled by intelligent ant colonies working over centuries (Central America). Locals believed the insects followed ancient instructions passed down from forgotten civilizations, building structures humans later claimed as their own.

The Immortal Cicadas Beneath the Earth
Some cultures believed cicadas never truly die but simply retreat underground to wait for the world to change (Eastern Asia). Their long, silent years beneath the soil were seen as proof they existed outside normal time, emerging only when conditions were right.

DID YOU KNOW?

Did you know? Some insects can **pause life without dying** by entering a state called *cryptobiosis* (world-wide). In this condition, their metabolism drops so low it's almost undetectable, allowing them to survive extreme cold, heat, dehydration, or lack of oxygen-conditions that would instantly kill most animals.

Did you know? Certain beetles and flies can **survive being frozen solid** (polar regions and high mountains). Instead of avoiding ice, their bodies produce special proteins that control how ice forms inside their cells, preventing damage and allowing them to thaw back to life when temperatures rise.

Did you know? Insects don't rely on lungs or blood to move oxygen the way humans do (world-wide). Oxygen travels directly through tiny tubes called **tracheae**, which means many insects can survive injuries or conditions that would stop breathing in other animals.

Did you know? Some insects can **repair massive DNA damage** caused by radiation far better than humans can (world-wide). This doesn't make them "radiation-proof," but it does mean their cells can recover from levels of exposure that would cause severe illness or death in mammals.

Did you know? A few insects can live **months without eating** by slowing their bodies to an extreme energy-saving mode (deserts and underground habitats). By conserving resources and recycling internal energy, they stretch survival far beyond what most life forms can manage.

STORY MOMENT

STORY MOMENT

Bugs That Break the Rules of Life

The night feels wrong. Not quiet-*paused*. Your flashlight slides across bark, frozen soil, and patches of snow when something twitches that absolutely should not. You stop cold. The insect in front of you is stiff, locked in ice like a tiny fossil. Dead. Then a leg bends.

Nearby, another insect crawls calmly across the snow, leaving thin tracks where nothing living should move. One hasn't breathed in minutes. Another shrugs off poison strong enough to kill animals hundreds of times its size. A third hasn't eaten in months-and doesn't seem bothered at all.

You step back, unsettled. These bugs don't struggle the way larger animals do. They don't panic. They shut down. They wait. Their bodies slow until time itself seems to ignore them. Some freeze solid and thaw later. Others lose body parts and keep going. A few survive being crushed, dried out, or starved, as if death is only a suggestion.

Your skin prickles as the idea sinks in. These aren't fragile creatures clinging to life. They're survivors built for disasters-fires, ice ages, poison, and collapse. When your light finally clicks off, the ground beneath you is still alive, crowded with tiny rule-breakers quietly doing what they've always done.

Tardigrade
Hypsibius dujardini
COMMON LOCATION: Worldwide. Survives extreme heat, freezing, radiation, and even the vacuum of space.

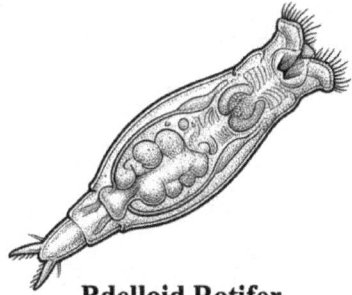

Bdelloid Rotifer
Adineta vaga
COMMON LOCATION: Freshwater environments worldwide. Can survive complete dehydration and reproduce without mating.

Bombardier Beetle
Brachinus spp.
COMMON LOCATION: Worldwide. Fires boiling chemical sprays to defend itself from predators.

Monarch Butterfly
Danaus plexippus
COMMON LOCATION: North America. Migrates thousands of miles across generations that never complete the full journey.

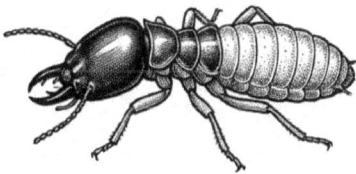

Termite
Isoptera
COMMON LOCATION: Tropical and subtropical regions worldwide. Builds massive climate-controlled structures without leaders or plans.

Caddisfly Larva
Trichoptera
COMMON LOCATION: Freshwater streams and rivers. Builds protective cases from sand, stone, shells, and debris.

FUN QUIZ

1. Which insect disables a cockroach by targeting the exact brain region responsible for escape behavior?

A) Tarantula hawk wasp

B) Emerald cockroach wasp

C) Assassin bug

D) Velvet ant

2. True or False: The American cockroach can survive long periods without breathing by closing its spiracles to conserve moisture and block toxins.

3. Which creature survives extreme Antarctic cold by using natural antifreeze that prevents ice from forming inside its cells?

A) Snow fly

B) Woolly bear caterpillar

C) Antarctic springtail

D) Ice worm

4. Which insect launches itself into the air using jaw snap recoil rather than legs?

A) Camel cricket

B) Trap-jaw ant

C) Jumping bristletail

D) Flea

5. True or False: The bombardier beetle stores boiling chemicals already mixed inside its body and releases them directly when threatened.

6. Which insect sees both above and below the water at the same time using divided eyes?

A) Water strider

B) Whirligig beetle

C) Backswimmer

D) Water boatman

7. True or False: The New Zealand glowworm attracts prey using glowing silk threads that function like a hanging light trap.

A mosquitoes life

QUIZ ANSWERS

1. **B**

2. **True**

3. **C**

4. **B**

5. **False** (the chemicals are stored separately and mixed only at firing)

6. **B**

7. **True**

9- The Strangest Fights in Nature

Insect battles, strategy, war, defense maneuvers

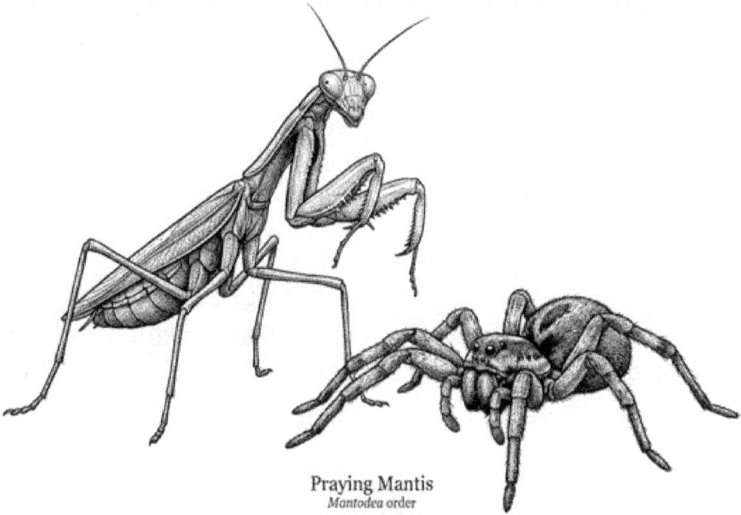

Praying Mantis
Mantodea order

Wolf Spider
Lycosidae family

COMMON LOCATION: Worldwide
A tense encounter between two roaming predators.

FUN & WEIRD FACTS

Bullet ants deliver one of the most excruciating stings on the planet. Found across the Amazon Basin, their venom overloads pain receptors so intensely that victims often compare the sensation to being shot.

Trap-jaw ants can fire themselves into the air faster than almost any animal movement recorded. These tropical insects live in forests worldwide, using mandibles that snap shut at over 100 mph to stun attackers or launch themselves away from danger.

The Japanese giant hornet can wipe out an entire beehive in under an hour. Native to East Asia, it uses slicing mandibles to decapitate defenders with frightening speed. Bees counter with the bizarre "heat-ball" tactic, smothering the hornet by cooking it alive.

Praying mantises often end fights before their rivals even register the attack. Found worldwide, they strike with forelegs so fast that the movement is nearly invisible. The spines on their arms lock opponents in place, making the outcome almost guaranteed.

Blue mud daubers raid spider webs with expert precision. Common in North America, they thread themselves between sticky strands and sting the spider before it can react. Even elaborate, well-built webs rarely slow them down.

Army ants overwhelm prey through sheer coordinated violence. Found in African and American tropics, their swarms behave like a single living machine. Anything that remains still for too long gets sliced apart within seconds.

Rhinoceros beetles hurl their rivals off logs with shocking strength. These beetles live in tropical regions worldwide and use their curved horns like natural grappling hooks, flipping opponents several feet during mating contests. Some falls end the fight instantly.

The velvet ant (a wingless wasp with an armor-hard shell) survives attacks that crush most insects. Found in North and South America, it endures repeated bites long enough to deliver one of the most painful stings in nature. Predators rarely attempt a second try.

Dragonflies dominate aerial battles with almost absurd accuracy. Found worldwide except Antarctica, they catch prey in midair using 360-degree vision. Their hunting success rate is so high that biologists consider them among the most efficient predators alive.

Harvester ants wage wars that leave miniature battlefields behind. Found across North and South America, thousands clash over territory in open fields. It's one of the few insect conflicts where casualties visibly shape the landscape.

Harvester Ants
Pogonomyrmex spp. (workers)

COMMON LOCATION:
Open fields, grasslands, and semi-arid regions across North and South America.

DESCRIPTION:
Harvester ants engage in large-scale territorial battles where thousands of individuals clash head-on. These conflicts can last for hours, leaving behind visible fields of fallen ants that mark contested boundaries. It is one of the rare insect wars where casualties are numerous enough to physically alter the landscape.

Trapdoor spiders win fights with explosive surprise attacks. Common in warm regions worldwide, they hide beneath a hinged soil lid and erupt upward the moment vibrations reach their burrow. Rivals often retreat after a single encounter.

The spitting spider (a small predator found in warm climates) defeats rivals by firing a zigzag spray of sticky silk that hardens instantly. This silk traps the opponent in place, giving the spider safe access for the finishing bite.

Tiger beetles run so fast they temporarily go blind mid-chase. Found worldwide, they must stop abruptly just to let their vision catch up. No other insect predator pushes its own speed limits this hard.

The devil's flower mantis (a brilliantly colored species from East Africa) uses camouflage and shock to win battles. It mimics a blooming flower until the moment it snaps forward. Rivals rarely anticipate the strike.

Termite soldiers block tunnels using their own oversized heads. Found in Africa, Asia, and Australia, these soldiers act as living barricades that prevent invading ants from entering the colony. Some even rupture defensive glands in a last act of protection.

Antlion larvae pull rivals into death traps of their own making. Found worldwide, they dig cone-shaped pits that collapse beneath struggling insects. Once prey begins sliding, escape becomes nearly impossible.

Horned treehoppers call in ant reinforcements when attacked. Native to Central and South America, they release signals that summon protective ant swarms. This alliance turns a fragile leaf-dweller into a well-defended partner.

The Jerusalem cricket (a large soil-dwelling insect from western North America) intimidates attackers by thumping its abdomen on the ground. These vibrations act as a warning signal that deters predators before the fight begins.

The feather-legged assassin bug (a spider-hunting species from Southeast Asia) hypnotizes rivals by waving silky hind legs. The motion mimics harmless debris, drawing spiders closer until the bug strikes with perfect timing.

FEATHER-LEGGED ASSASSIN BUG
Ptilocnemus lemur (adult)

COMMON LOCATION:
Rainforests and woodland edges of Southeast Asia.

DESCRIPTION:
The feather-legged assassin bug hunts spiders by slowly waving its silky hind legs. The motion resembles drifting debris, luring spiders closer instead of alarming them. When the prey commits, the bug strikes with precise timing, turning deception into a lethal advantage.

Driver ants can overpower animals hundreds of times their size. Found in central and eastern Africa, their jaws lock onto skin so tightly that removal often requires tools. A single column can strip a carcass clean in minutes.

Rove beetles win many fights simply by bluffing. Found worldwide, they raise their abdomen like a scorpion despite lacking a stinger. Many predators flee before discovering the trick.

Pseudoscorpions (tiny bark-dwelling arachnids found worldwide) use venom-tipped pincers in microscopic battles. They overpower mites and springtails with surprising force. Despite their size, they're capable and relentless predators.

The spiny flower mantis startles enemies with sudden flashes of color. Native to Africa, it flares its patterned wings like a warning flag just before striking. The display buys milliseconds that often decide the fight.

Woolly aphids summon ant bodyguards when threatened. Found in North America, Europe, and Asia, they release distress chemicals that bring ant reinforcements within seconds. These alliances protect otherwise helpless insects.

Scorpionflies win confrontations by pretending to be dangerous. Found worldwide, they raise their harmless tails like stingers. Predators often retreat immediately, fooled by the bluff.

Gladiator spiders cast expanding nets onto rivals in a single spring-loaded strike. Found in Australia and tropical regions, they stretch a web between their legs and launch it over the opponent. The net tightens instantly like a weighted blanket.

The emerald cockroach wasp removes the "escape reflex" from its victim. Native to tropical Asia, it stings the cockroach's brain in the precise spot that triggers fleeing. The roach remains alive but disturbingly cooperative.

Velvet worms immobilize rivals using high-pressure jets of adhesive slime. Found in Australia, New Zealand, and tropical forests worldwide, the glue hardens instantly. It's one of the strangest battle tools in the animal kingdom.

Net-winged beetles rely on toxic colors instead of combat. Found in Southeast Asia and the Americas, their bright orange bodies warn predators of danger. One bad bite is usually enough to prevent future attacks.

The feather-horned beetle (a rare species from Australia) senses rivals with its ornate antennae. These feather-like structures detect subtle air currents during confrontations. Battles are often decided by timing rather than strength.

Diving beetles overpower rivals underwater using burst-speed strikes. Found worldwide, they cut through prey with razor-sharp jaws. Their aquatic agility makes them apex predators in small ponds.

Ant-mimicking spiders avoid fights by pretending to be ants. Found in tropical regions globally, predators hesitate to attack them for fear of ant retaliation. Their disguise becomes a powerful survival weapon.

The woolly bear caterpillar survives attacks by curling into a bristly armored coil. Found across North America and Europe, its hairs irritate predators that try to bite through. Birds quickly learn to avoid them.

Woolly Bear Caterpillar
Pyrrharctia isabella (larval stage)

COMMON LOCATION:
Widely distributed across North America and parts of Europe.

DESCRIPTION:
The woolly bear caterpillar defends itself by curling into a dense, bristly coil when threatened. Its stiff hairs irritate the mouths and faces of predators that attempt to bite through. Birds quickly learn to recognize the caterpillar's texture and avoid it after painful encounters.

Lanternfly nymphs escape predators with chaotic, explosive jumps. Native to Asia but now spreading worldwide, their unpredictable movements confuse even skilled hunters.

Fire ants coordinate painful attacks by locking onto skin with their jaws. Native to South America and invasive in North America, they hold position so they can sting repeatedly. The coordinated assault magnifies the pain dramatically.

The monarch caterpillar turns milkweed toxins into a chemical shield. Found across North America, it stores poisons that make predators vomit shortly after eating one. Most attackers never attempt a second try.

Spiny orb-weaver spiders avoid becoming bird snacks thanks to their armor-like spikes. Found worldwide in warm regions, the spikes make swallowing uncomfortable and risky. Many birds release them mid-attack.

The trapdoor ant (a rarely seen underground hunter from Southeast Asia) bursts upward from soil chambers with sudden force. Rivals barely react before being grabbed.

Assassin bugs liquefy their enemies from the inside. Found worldwide, their beak injects enzymes that dissolve tissue. The bug then drinks the liquefied meal.

Kissing bugs avoid being eaten simply because predators know they carry dangerous parasites. Found in the Americas, this reputation alone deters attackers. Sometimes survival is a matter of branding.

Kissing Bug
Triatoma spp. (adult)

COMMON LOCATION:
Widespread across Central and South America, with some species extending into southern North America.

DESCRIPTION:
Kissing bugs are avoided by many predators because they are known carriers of dangerous parasites. This reputation alone provides protection, reducing attacks before they happen. In this case, survival depends not on strength or speed, but on a well-earned warning.

Camel crickets escape danger with enormous leaps in total darkness. Common in North America, their chaotic bouncing disorients predators. It's a simple defense, but shockingly effective.

Bombardier midge larvae fire tiny pulses of water at attackers. Found in North American streams, the jets disrupt surface tension and create an escape window. It's a miniature version of the bombardier beetle's defense.

Soldier aphids willingly sacrifice themselves to protect their colony. Found worldwide on trees and crops, they plug holes in plant tissue with their own bodies. This living cork stops predators from entering.

Hercules beetles lift rivals many times their own weight. Native to Central and South America, they use their massive horns to flip opponents off branches. Their strength-to-size ratio is among the greatest in nature.

Net-casting spiders immobilize prey larger than themselves using expanding webs. Found in Australia, Africa, and the Americas, the web unfolds over a victim like a weighted net. Once wrapped, escape is nearly impossible.

What is Chagas disease?

Chagas disease, also known as American trypanosomiasis, is a potentially life-threatening disease caused by the parasite *Trypanosoma cruzi*. It **is most commonly transmitted by biting insects known as 'kissing bugs'** that are infected with the parasite. It can also be passed from mother to child during pregnancy and childbirth. As people typically show no symptoms for years, most are unaware they have Chagas.

Mantis vs Spider
Mantis religiosa vs Araneae
COMMON LOCATION: Worldwide.
A praying mantis ambushes a spider, reversing the usual predator–prey roles.

Ant vs Termite
Formicidae vs Isoptera
COMMON LOCATION: Tropical and subtropical regions.
Organized ant raiders attack termite colonies in coordinated assaults.

Stag Beetle Duel
Lucanus cervus
COMMON LOCATION: Europe and Asia.
Males wrestle using oversized jaws to throw rivals from branches.

Spider Wasp vs Spider
Pompilidae vs Araneae
COMMON LOCATION: Worldwide.
A wasp delivers a precise sting to paralyze a spider for its larvae.

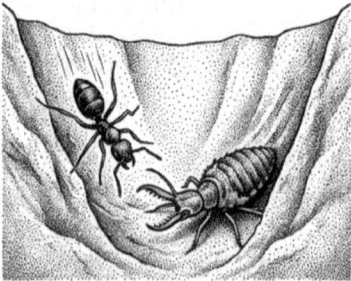

Antlion vs Ant
Myrmeleontidae vs Formicidae
COMMON LOCATION: Sandy regions worldwide.
An antlion larva drags prey into a sand pit using collapsing walls.

Honeybee Defense Ball
Apis mellifera
COMMON LOCATION: Worldwide.
Worker bees swarm an intruder, overheating it to stop the threat.

MYTHS - BUSTED

Myth: All bees die after they sting an enemy. Actually, most stinging insects are "reusable weapons." Only the honeybee has a barbed stinger that gets stuck in thick skin (like mammals), ripping out the bee's internal organs when it flies away. Bumblebees, wasps, hornets, and yellow jackets have smooth stingers, meaning they can sting an enemy repeatedly during a battle and fly away to fight another day.

Myth: Female Praying Mantises always eat the male's head after a "date." While mantises are famous for being the femme fatales of the insect world, they don't *always* choose violence. Studies show that females only eat the males about 13% to 28% of the time in the wild. This usually only happens if the female is starving or incredibly stressed. If she is well-fed, the male usually escapes with his head attached!

Myth: Daddy Longlegs have the world's most deadly venom, but their mouths are too small to bite humans. This is an urban legend that refuses to die. First, "Daddy Longlegs" can refer to harvestmen (which aren't true spiders and have no venom at all) or cellar spiders. While cellar spiders do have venom, it is very weak-designed for tiny insects, not people. It certainly isn't the "most deadly," and it wouldn't hurt you even if they did manage to bite.

Myth: Scorpions will sting themselves to death if surrounded by fire. It might look like a scorpion is committing suicide when trapped near heat, but it's actually a biological glitch. Scorpions are cold-blooded and extremely sensitive to heat. When they get too hot, their bodies spasm uncontrollably, causing their tail to strike wildly. They aren't giving up the fight; they are just overheating!

Myth: The biggest bug always wins the fight. In the insect world, chemical warfare often beats raw strength. The Bombardier Beetle, for example, is small, but it can defeat much larger predators by shooting a boiling hot, toxic chemical spray from its rear end with a loud *POP*. Size doesn't matter when you are a walking flamethrower.

Myth: Earwigs use their pincers to pinch your brain through your ear. Earwigs have zero interest in your brain. While they do like dark, tight spaces, their pincers (cerci) are actually weapons of war used against *other* insects. They use them to wrestle rivals, capture prey, or unfold their own wings. If an earwig pinches a human, it's just a mild defensive nip, not a brain invasion.

Myth: Termites are just defenseless wood-eaters. Termites are actually highly trained soldiers involved in an ancient, never-ending war with ants. Termite colonies have a "caste" system that includes soldier termites with massive, specialized heads. Some have jaws so large they can't feed themselves, while others (Nasute termites) have heads shaped like nozzles that shoot sticky "glue" to trap invading ants.

Myth: Centipedes bite their enemies with their jaws. Technically, centipedes don't bite; they pinch-stab. They don't use their mouthparts to deliver venom. Instead, their first pair of legs has evolved into sharp, hollow claws called *forcipules*. They use these legs to pierce the armor of their enemies and inject venom like a hypodermic needle.

LEGENDS

LEGENDS

The Devil's Darning Needle The Sewing Assassin - Europe & North America For centuries, children were told to keep their mouths shut and eyes open when walking near streams. Folklore warned that if a child told a lie-or fell asleep by the water-a long, slender creature would swoop down and sew their lips or eyelids shut with its body. In reality, dragonflies are harmless to humans. They are actually the fighter pilots of the insect world, using their legs not as needles, but as a "basket" to snatch other flying insects out of the air in brutal mid-flight dogfights.

The Prophet of Doom The Praying Mantis - Southern Europe In rural France and Italy, it was once believed that if you were lost, the Praying Mantis would gently point a limb to guide you home. Others believed its posture was a sign of deep religious devotion. The truth is far more violent. The mantis is not praying; it is assuming a "raptorial" combat stance. It holds its spiked forelegs in that specific position to generate maximum leverage for a lightning-fast ambush strike, capable of spearing passing prey in milliseconds.

The Thunder Carrier Stag Beetles - Germany German folklore connected the massive Stag Beetle to Thor, the god of thunder. Peasants believed these insects summoned lightning storms and carried burning coals from the underworld into thatched roofs to start fires. Because of this, people were terrified to touch them. Today we know those terrifying "horns" have nothing to do with lightning. They are actually oversized jaws used exclusively for wrestling matches. Male beetles lock jaws and try to suplex each other off tree branches to win the right to mate.

The Dancing Plague Tarantism - Italy In the 17th century, villagers believed that the bite of the Wolf Spider (mistakenly called a Tarantula) was fatal unless the victim engaged in frantic, non-stop combat with the poison. This led to "Tarantism," where victims would dance maniacally for days to sweat out the venom. Science has since proven the spider's venom is mild, similar to a bee sting. The "fight" against the venom was likely mass hysteria-or simply a convenient excuse for people to party and dance during strict religious times.

The Brain Borer - The Earwig (England)

Few insects have a reputation as terrifying as the earwig. The legend, dating back nearly a thousand years, claims it crawls into sleeping humans' ears, bores into the brain, and lays eggs that cause madness. While earwigs do prefer dark, tight spaces, this is a complete myth. The rear pinchers (cerci) aren't for brain invasion-they're defensive tools used to deter predators or battle rival earwigs.

The Omen of War - The 17-Year Cicada (North America)

When periodic cicadas emerged in their billions, early American settlers were horrified by a distinct orange "W" shape on their wings, believing it signaled coming war. Communities panicked and prepared for conflict whenever the insects appeared. The mark is simply a convergence of stiff wing veins for flight. The only "war" cicadas fight is **predator satiation**-overwhelming enemies with sheer numbers to survive.

The Samurai Soul Heikegani Crabs - Japan *Note: While technically a crustacean, this fits the "Tiny Monster" theme perfectly.* Fishermen in Japan often throw back crabs with shells that look like angry human faces. Legend says these crabs contain the souls of the Heike samurai warriors who died in a massive naval battle in 1185, forever scowling in defeat. In a twist of accidental evolution, this is a defensive strategy driven by humans. By refusing to eat crabs that looked like faces, humans selectively bred the crabs to look more and more like angry warriors over centuries, protecting them from the dinner pot.

DID YOU KNOW?

Did you know? The Bombardier Beetle carries a chemical weapon in its abdomen that is essentially a biological machine gun. When threatened, it mixes chemicals to create a boiling hot, toxic spray that shoots out at 212°F (100°C) with a loud popping sound, scalding predators instantly.

Did you know? Japanese Honeybees have developed a "cooking" defense against their arch-enemy, the Giant Hornet. When a scout hornet enters the hive, hundreds of bees ball around it and vibrate their muscles to raise the temperature to exactly 117°F. This heats the ball enough to roast the hornet alive, but is just cool enough for the bees to survive.

Did you know? The Trap-Jaw Ant possesses the fastest strike in the animal kingdom, snapping its mandibles shut at speeds of up to 145 miles per hour. The bite is so powerful that if the ant snaps against the ground, the recoil launches the ant backwards through the air, acting like an emergency ejection seat to escape danger.

Did you know? Older workers of the termite species *Neocapritermes taracua* develop blue crystals on their backs that act as a suicide vest. When an enemy invades, these aging termites rupture their own bodies, mixing the crystals with internal fluids to create a toxic, sticky blue explosion that kills both them and the attacker.

Did you know? The Assassin Bug is the master of stealth warfare, wearing the corpses of its victims as camouflage. After draining the fluids from ants, it stacks the hollow exoskeletons on its back to create a gruesome armor that hides its scent and confuses larger predators like spiders.

Did you know? The Diabolical Ironclad Beetle has armor so tough it can survive being run over by a car. Its exoskeleton is fused together like a complex jigsaw puzzle, allowing it to withstand force 39,000 times its own body weight. The shell is so hard that entomologists often bend their steel pins trying to mount them.

Did you know? Male Stalk-Eyed Flies avoid physical violence entirely by having "measuring contests." They stand face-to-face and compare the width of their eye stalks; the fly with the wider span is declared the winner and takes the territory, proving that sometimes sizing up your opponent is better than fighting them.

Did you know? The Hawk Moth Caterpillar defends itself by shapeshifting into a deadly pit viper. When threatened, it puffs up its lower body, which features markings that look exactly like snake eyes and scales. It even mimics the snake's striking motion to terrify birds that would otherwise eat it.

Did you know? The Green Lacewing larva is a "wolf in sheep's clothing" that infiltrates woolly aphid colonies to eat them. To avoid detection by the ant guards protecting the aphids, the larva plucks the waxy white fluff off the aphids and glues it to its own back, becoming invisible to the security forces.

Did you know? Slave-Making Ants are incapable of feeding themselves, so they rely entirely on war. They launch organized raids on the nests of other ant species, using panic pheromones to scatter the defenders, and steal the pupae (cocoons). When the stolen ants hatch, they work as slaves, believing the raiders are their true family.

STORY MOMENT

STORY MOMENT

The River of Teeth

The jungle usually hums with life, but then-silence. The birds stop singing. The crickets go quiet. Even the air feels tense, like it's waiting. A dry rustling sound creeps out of the underbrush, the kind you'd expect from heavy rain hitting brittle leaves-except the sky above you is perfectly clear. You look down, and your stomach tightens. The ground is moving.

A massive river of army ants, millions strong, pours across roots and rocks like a living shadow. They aren't wandering. They're hunting. Ahead of them, the forest explodes with panic. Spiders scramble upward. Beetles race for cover. Scorpions flee in every direction, driven by an instinct that screams *run*. The ants keep coming.

They move with eerie precision. The column splits and rejoins without slowing. Gaps disappear as ants lock together, turning their own bodies into bridges. No one ant is in charge, yet the swarm acts with perfect coordination, like a single mind with one goal.

You step back as the river flows past your feet. When it finally moves on, the forest floor is strangely empty and silent, stripped clean in its path. That's when it sinks in-you haven't just watched insects at work. You've just witnessed one of the most organized and unstoppable forces in nature.

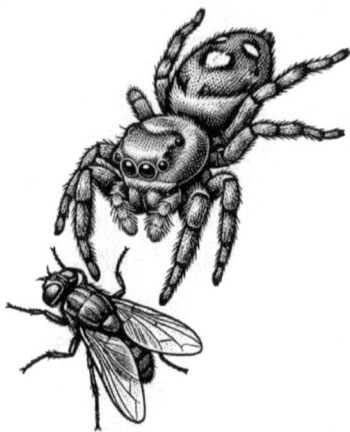

Jumping Spider vs Fly
Salticidae vs Musca domestica
COMMON LOCATION: Worldwide.
A jumping spider launches a precision
leap to capture prey in mid-motion.

Robber Fly Ambush
Asilidae
COMMON LOCATION: Worldwide.
A robber fly intercepts flying insects
mid-air using speed and grip.

Dragonfly Intercept
Anisoptera
COMMON LOCATION: Near freshwater
worldwide. A dragonfly captures prey
mid-flight using basket-like legs.

Velvet Ant Defense
Mutillidae
COMMON LOCATION: Warm regions worldwide.
A wingless wasp resists attack with an extremely
painful sting and armored body.

FUN QUIZ

1. How do Japanese honeybees defeat a giant hornet that invades their hive? A) They sting it simultaneously until it runs away. B) They swarm around it and vibrate to cook it with body heat. C) They cover it in sticky wax so it cannot fly. D) They use a chemical spray to blind it.

2. The Trap-jaw ant is famous for having mandibles (jaws) that can do what? A) Snap shut at over 100 mph. B) Glow in the dark to scare predators. C) Inject a venom that puts enemies to sleep. D) Chew through solid stone.

3. Why is the "Bullet Ant" given its terrifying name? A) It is shaped exactly like a metal bullet. B) It can shoot tiny projectiles from its tail. C) Its sting is so painful that victims compare it to being shot. D) It moves faster than a speeding bullet.

4. How does the Emerald Cockroach Wasp defeat a cockroach much larger than itself? A) It spins a web around the cockroach. B) It calls thousands of other wasps to help attack. C) It stings the cockroach's brain to remove its "escape reflex." D) It bites off the cockroach's legs so it cannot run.

5. Soldiers of certain termite species defend their colony using which strange body part? A) Their oversized heads, which plug tunnels like a cork. B) Their long tails, which act like whips. C) Their wings, which create a wind to blow ants away. D) Their legs, which have sharp spikes for kicking.

QUIZ ANSWERS

1. **B** (They swarm around, vibrate to cook it with body heat.)

2. **A** (Snap shut at over 100 mph.)

3. **C** (sting is so painful that victims compare it to being shot.)

4. **C** (stings the cockroach's brain to remove its "escape reflex.")

5. **A** (Their oversized heads, which plug tunnels like a cork.)

Bug Smiles

Q. How do bees brush their hair?

A. With a honeycomb.

Q. Why don't insects use social media?

A. Too many *bugs* in the system.

Q. Why was the bug carrying the bottle of air freshener?

A. Because it had a little bug odor.

Q. Why did the principal hate insect jokes?

A. Because they really bugged him.

Q. Why did the termite bring a suitcase to class?

A. It heard there was a lot of *wood*work involved.

Q. Why don't mosquitoes ever win arguments?

A. They always *suck* at making their point.

Q. Where do wasps go when they get sick?

A. To the wasp...ital.

Q. What do fireflies eat between meals?

A. A light snack.

Q. Why did the spider refuse to answer emails?

A. Too many *attachments*.

Q. Why wasn't the butterfly invited to the dance?

A. Because it was a wallflower.

Q. Why are frogs so happy?

A. Because they eat whatever bugs them.

CHAPTER 10

10- Insects in Human Fears, Myths & Nighttime Legends

More myths, ancient legends, nighttime superstitions

"Tarantula Hawk Wasp
Pepsis spp.

COMMON LOCATION: Americas
Delivers one of the most painful insect stings ever recorded.

Tarantula
Theraphosidae family

COMMON LOCATION: Worldwide
Uses venomous fangs and powerful legs to overpower prey.

FUN & WEIRD FACTS

MIND-BLOWN MOMENT

Fear of spiders may be wired into the human brain before we can even speak. Studies show babies react faster to images of **spiders** (eight-legged predators found worldwide) than to flowers or faces. This suggests our ancestors survived by noticing danger first-and thinking later.

Scorpions glow ghostly blue under ultraviolet light, turning deserts into haunted landscapes after dark. The eerie shine of **scorpions** (venomous arachnids found on every continent except Antarctica) comes from chemicals in their exoskeleton. For centuries, glowing scorpions fueled myths of cursed sands and spirit creatures.

Moths symbolize spirits in nighttime folklore around the world. A pale **moth** (nocturnal flying insect found globally) entering a home was believed to be a visiting soul. Their silent flight and attraction to flame made them messengers between worlds.

The Black Widow earned a reputation far darker than its behavior deserves. Legends paint **Black Widow spiders** (venomous spiders from the Americas) as aggressive killers, yet they usually freeze or play dead. Most bites only occur when humans accidentally press them against skin.

Daddy Longlegs myths claim deadly venom trapped behind tiny fangs. In reality, **Daddy Longlegs** (arachnids found worldwide) are either harvestmen with no venom or cellar spiders with weak venom harmless to humans. Playground fear turned harmless legs into legends.

Bed bugs awaken only when darkness signals safety. **Bed bugs** (blood-feeding insects found worldwide in furniture) sense carbon dioxide from human breath as an invisible alarm clock. Nighttime feeding helped turn them into symbols of unseen invasion.

Cockroaches survive screams, stomps, and sudden chaos- sometimes by mistake. **Cockroaches** (scavenging insects found worldwide) navigate using air currents, not sound. Loud movement can confuse them into sprinting toward the threat, deepening fear.

The Deathwatch Beetle tapped fear into wooden walls. The ticking call of **Deathwatch Beetles** (wood-boring beetles from old European buildings) sounded like a countdown to death. In truth, it was only a love call echoing through beams.

DEATHWATCH BEETLE
Xestobium rufovillosum (adult)

COMMON LOCATION:
Old wooden buildings, beams, and structural timbers across Europe.

DESCRIPTION:
The deathwatch beetle earned its name from a sharp tapping sound heard in quiet wooden rooms, once believed to foretell death. The rhythmic ticking, echoing through old beams, sounded like a grim countdown to those who heard it at night. In reality, the beetle was alive and active, striking the wood to attract a mate—its love call amplified by centuries-old timbers.

Dragonflies became demons in old European myths. Nicknamed Devil's Darning Needles, **dragonflies** (predatory flying insects found near water worldwide) were said to sew shut lying mouths. The story scared children into honesty, not silence.

Scarab beetles shaped gods in ancient Egypt. The rolling motion of **scarab beetles** (dung beetles from North Africa) inspired myths of the sun being pushed across the sky. Beetles became symbols of rebirth and eternal cycles.

Camel spiders terrified soldiers through rumor alone. Photos exaggerated **Camel Spiders** (large desert arachnids from the

Middle East) into screaming giants. In reality, they chase shadows to escape heat, not people.

Fireflies sparked legends of wandering souls. The glowing pulses of **fireflies** (bioluminescent beetles found worldwide) were believed to be spirits drifting between lives. Their cold light made darkness feel alive.

Wolf spiders explode fear when disturbed. Female **Wolf Spiders** (ground-hunting spiders found worldwide) carry hundreds of babies on their backs. When startled, the swarm scatters in all directions, shocking observers.

Ticks hunt silently in grass like tiny assassins. **Ticks** (parasitic arachnids found worldwide) wait with legs raised in a behavior called questing. They latch on without being felt, feeding fear of the unseen.

The Manticore was born from insect terror. Ancient stories fused lions with **scorpion** tails (venomous arachnids known across deserts). A small sting became a monster that devoured men.

Silverfish move wrong for land creatures. The motion of **silverfish** (primitive insects found in dark homes) resembles swimming, not walking. Their slippery movement unsettles people instinctively.

Wasps remember faces long after encounters end. **Golden Paper Wasps** (social wasps from warm climates) can recognize individuals. Swatting one may make you memorable.

Cicada shells haunt trees like ghosts. When **cicadas** (loud insects from temperate regions) molt, they leave hollow replicas behind. Finding one staring back can feel unsettlingly alive.

Trypophobia mirrors nests built by insects. The fear of clustered holes resembles the patterns of **bee** and **wasp** nests (stinging insects found worldwide). The brain reacts before logic intervenes.

Africanized Honey Bees redefined fear through numbers. These **bees** (hybrid honey bees from the Americas) swarm aggressively when threatened. Their collective defense fueled modern legends of unstoppable insects.

Africanized Honey Bees
Apis mellifera scutellata hybrids (workers)

COMMON LOCATION:
Widespread across Central and South America, with established populations in parts of North America.

DESCRIPTION:
Africanized honey bees transformed fear by sheer numbers rather than size. When disturbed, thousands respond at once, pursuing threats far beyond the nest. This overwhelming collective defense reshaped modern legends of insects that cannot be outrun or reasoned with.

House centipedes race like living shadows. **House Centipedes** (fast predators found worldwide indoors) move with blinding speed on fifteen leg pairs. Fear follows motion faster than thought.

Coffin Flies breach the boundary of burial. **Coffin Flies** (small flies associated with decay) can reach buried remains through soil. Their behavior shaped gothic nightmares.

Bullet Ants define pain rituals. The sting of **Bullet Ants** (large rainforest ants from the Amazon) is so intense it fuels rites of passage. Fear becomes proof of strength.

Praying mantises stare like thinking beings. **Praying Mantises** (predatory insects found worldwide) rotate their heads nearly 180 degrees. Eye contact unnerves predators and humans alike.

Locust swarms resemble apocalypse visions. **Desert Locusts** (grasshoppers from Africa and Asia) can darken skies and erase crops. Ancient texts immortalized their devastation.

Saddleback caterpillars sting like living traps. The bright colors of Saddleback Caterpillars (moth larvae from North America) warn predators too late. Pain teaches respect instantly.

Zombies exist in insect reality. A fungus controls **ants** (social insects from tropical forests), forcing fatal behavior. Nature wrote horror long before fiction.

Goliath Birdeaters spiders loom larger than fear. Despite the name, Goliath Birdeater spiders (giant tarantulas from South America) rarely eat birds. Size alone carries legend.

Goliath Birdeater Spider
Theraphosa blondi (adult)

COMMON LOCATION:
Tropical rainforests of northern South America.

DESCRIPTION:
The Goliath birdeater looms larger than fear itself, its sheer size enough to inspire legend. Despite the name, these massive tarantulas rarely prey on birds, feeding instead on insects and small animals. Size alone—legs spread wide and body heavy—turned a real spider into a creature of myth.

Crane Flies die for mistaken identity. **Crane Flies** (harmless insects found worldwide) are killed as giant mosquitoes. Fear ignores facts when wings look wrong.

Termites eat homes while owners sleep. **Termites** (wood-eating insects found globally) work nonstop, unseen. The terror is silent loss, not bites.

Mites live quietly on human faces. **Demodex mites** (microscopic arachnids in human pores) are harmless roommates. Knowing they exist is enough to unsettle sleep.

Firebrats love heat humans flee. **Firebrats** (heat-loving insects found indoors) thrive near ovens and boilers. Their survival defies comfort rules.

Assassin bugs melt prey from the inside. **Assassin Bugs** (predatory insects found worldwide) inject enzymes before feeding. Liquid meals inspired fearsome names.

Antlions build traps worthy of legends. The pits of **antlion larvae** (sand-dwelling predators found globally) collapse beneath victims. Perfect geometry becomes doom.

Backswimmers hunt upside-down beneath moonlit water. Backswimmers (aquatic insects found worldwide) watch silhouettes above. The world flips at night.

Water striders defy surface tension myths. **Water Striders** (pond insects found worldwide) skate effortlessly on water. Physics feels broken after dark.

Glowworms create stars underground. **New Zealand Glowworms** (bioluminescent larvae from caves) turn ceilings into constellations. Beauty lures prey-and fear.

Earwigs earned names through superstition alone. **Earwigs** (nocturnal insects with pincers found worldwide) do not invade brains. Darkness invented danger.

Mosquitoes kill silently through disease. **Mosquitoes** (blood-feeding flies found worldwide) have caused more human deaths than wars. Nighttime buzzing signals real threat.

Huntsman spiders teleport in perception only. The speed of Huntsman spiders (fast arachnids from Australia and Asia) tricks the eye. Fear fills the gaps.

MYTHS - BUSTED

MYTH: The average person swallows eight spiders a year while they sleep.

This is one of the most famous false facts in history. It was created by a journalist in 1993 to show how quickly fake information spreads. In reality, a sleeping person's mouth is a terrifying place for a spider-it's windy, wet, and noisy from breathing or snoring. Spiders sense danger through vibrations, and a human feels like a shaking volcano. They avoid us completely.

MYTH: Daddy Longlegs are the most poisonous spiders on Earth, but their fangs are too short to bite humans.

This legend is wrong on multiple levels. "Daddy Longlegs" usually refers to crane flies (which are harmless) or harvestmen (which have no venom at all). Even true cellar spiders-sometimes called Daddy Longlegs-have very weak venom that causes nothing more than mild irritation.

MYTH: Earwigs get their name because they burrow into human ears and lay eggs in the brain.

While the name is creepy, the insect is innocent. The word "earwig" comes from Old English *ear-wicga* ("ear wiggler"), likely because they hide in tight, dark crevices. Even if one did wander into an ear-which is extremely rare-it cannot bore through the eardrum or reach the brain. Earwigs prefer damp leaves and rotting wood, not gray matter.

MYTH: Cockroaches are the only creatures that will survive a nuclear bomb.

Cockroaches are tough and can tolerate far more radiation than humans, but they are not invincible. A nuclear blast would still kill them instantly. In fact, some insects-such as fruit flies and certain parasitic wasps-can survive higher radiation doses than cockroaches can.

MYTH: Mosquitoes bite you more because you have "sweet blood."

Mosquitoes aren't attracted to sugar; they're drawn to smell and breath. They seek carbon dioxide, body heat, and chemicals like lactic acid. People who exhale more CO_2, have Type O blood, or produce certain skin chemicals are more likely to be bitten.

MYTH: Camel spiders are half the size of a human and scream as they run.

Photos from the Middle East made these arachnids look enormous due to forced perspective, but in reality they grow to about six inches at most. They don't scream-some make a faint hissing sound. When they appear to chase people, they're actually running toward shade to escape the desert heat.

MYTH: You only get bed bugs if your house is dirty.

Bed bugs don't care about mess or cleanliness-they care about blood. They're just as common in luxury hotels as in cluttered homes. These insects hitchhike on luggage, clothing, and furniture, meaning anyone can bring them home.

MYTH: Bees always die after they sting you.

This is only true for honey bees, whose barbed stingers get stuck in human skin. Most other stinging insects-such as bumblebees, wasps, hornets, and yellow jackets-have smooth stingers and can sting repeatedly without dying.

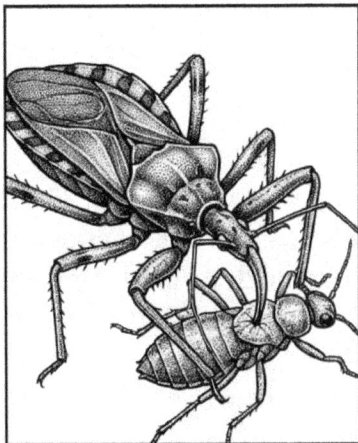

Assassin Bug
Reduviidae

COMMON LOCATION: Worldwide.
A stealthy predator that pierces prey with a needle-like mouthpart, inspiring fear through its sudden attacks.

Saddleback Caterpillar
Acharia stimulea

COMMON LOCATION: Eastern North America.
Bright warning colors hide venomous spines that cause painful stings when touched.

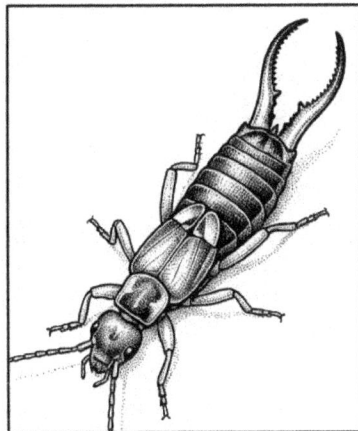

Earwig
Dermaptera

COMMON LOCATION: Worldwide.
Long surrounded by myths about crawling into ears, despite being mostly harmless scavengers.

Mite
Acari

COMMON LOCATION: Worldwide.
Nearly invisible creatures linked to itching, infestations, and long-standing human discomfort.

The Skull Bearer The Death's Head Hawkmoth - Europe For centuries, finding this large moth on a doorstep was considered a guaranteed omen of death for the household. Its reputation was fueled by the distinct human skull pattern on its back and its ability to emit a loud, mouse-like squeak when disturbed-thought to be the scream of a tormented soul. In reality, the "scream" is a tool used to mimic the scent and sound of a queen bee, allowing the moth to sneak into beehives and steal honey without being attacked.

The Devil's Steed The Devil's Coach Horse - Great Britain In the Middle Ages, this black beetle was feared as a familiar of Satan. Folklore claimed that if it pointed its curled tail at you, you were cursed, and if you killed one, the seven deadly sins would be forgiven. The beetle's terrifying posture-curling its abdomen over its head like a scorpion-is actually just a bluff. It has no stinger and no venom; it simply mimics a dangerous creature to scare away birds and curious humans.

The Shadow of Death The Black Witch Moth - Mexico & The Caribbean Known in folklore as the "Mariposa de la Muerte," this bat-sized moth is one of the most feared insects in the Western Hemisphere. Legend says that if it flies into a bedroom and lands above the bed, the person sleeping there will not wake up. Despite the terror it inspires, the moth is completely harmless. It is simply a fruit-eater that often seeks shelter in cool, dark houses to rest during its long migration.

The Witch's Curse Cuckoo Spit - Worldwide Walking through meadows, children were often warned never to touch the clumps of white foam found on plant stems. Elders claimed this was "witch's spit" or "snake spit," and touching it would cause blindness or bad luck. The foam is actually an ingenious survival house built by Spittlebug nymphs. They excrete the fluid and pump air into it to create a bubble fortress that hides them from predators and keeps their soft bodies from drying out.

The Wandering Souls Fireflies - Japan In Japanese folklore, the floating, flickering lights of fireflies (*Hotaru*) were not seen as insects, but as the souls of dead warriors or ancestors lingering in the mortal world. Catching them was sometimes seen as disturbing the dead. Science reveals a more romantic truth: the lights are a complex language of love. Each species flashes a specific Morse-code pattern to identify themselves to potential mates in the dark grass.

The Thunder Caller Thunderbugs (Thrips) - Europe Farmers once believed that tiny, itching swarms of black insects were supernatural summoners of summer storms. They appeared seemingly out of thin air right before thunder cracked, leading people to believe they brought the lightning. We now know that these insects, called Thrips, are just incredibly sensitive to atmospheric pressure. When air pressure drops before a storm, they lose the ability to fly high and are forced down to ground level in massive, irritating clouds.

The Grave Digger The Burying Beetle - North America Travelers in the woods often reported seeing the earth move on its own, swallowing dead birds or mice whole in a single night. Superstition held that invisible spirits or "earth demons" were dragging the bodies to hell. The culprit is the hardworking Burying Beetle. A pair of beetles will dig the soil out from under a carcass to sink it underground, preserving the meat to feed their larvae in a safe, subterranean nursery.

DID YOU KNOW?

Did you know? A cockroach can live for weeks without its head. Because they breathe through tiny holes in their body segments (spiracles) and have a decentralized nervous system, a decapitated cockroach doesn't suffocate. It eventually dies only because it has no mouth to drink water.

Did you know? There are moths in the Amazon that drink the tears of sleeping animals. These "vampire moths" use a specialized proboscis (straw) to irritate the eyelids of sleeping birds-and sometimes humans-to sip the tear fluid, which provides them with vital salt and proteins.

Did you know? Scorpions are practically indestructible and can survive being frozen solid. You can freeze a scorpion in a block of ice overnight, let it thaw out in the sun, and it will walk away as if nothing happened. They can also survive underwater for 48 hours and go a full year without eating.

Did you know? The annoying high-pitched whine of a mosquito is actually a love song. The buzzing sound is created by their wings beating 300 to 600 times per second. Males and females adjust their pitch to harmonize with each other in mid-air to find a mate in the dark.

Did you know? Blowflies are often the first "detectives" to arrive at a crime scene. They can smell organic decay from over a mile away and will land on a body within minutes. Forensic scientists use the age of the maggots to determine the exact time of death, solving murders that humans can't.

Did you know? Bed bugs hunted dinosaurs long before they hunted us. Fossil evidence suggests that the ancestors of modern bed bugs were scurrying around 100 million years ago. They have survived the asteroid that killed the T-Rex, proving they are one of the ultimate survivors in history.

Did you know? Your mattress might be a zoo for millions of microscopic creatures. The average used mattress can house anywhere from 100,000 to 10 million dust mites. The reason people are allergic to them isn't the mite itself, but a protein found in their waste pellets.

Did you know? Silverfish are older than the dinosaurs. These alien-looking silver insects have been crawling on Earth for 400 million years. They survived the mass extinction events that wiped out the dinosaurs and the ice ages, remaining virtually unchanged since prehistoric times.

Did you know? Some tarantulas keep tiny frogs as pets. In a strange alliance, the Colombian Lesser black Tarantula protects the dotted humming frog from predators. In return, the tiny frog eats the ants and flies that would otherwise attack the spider's eggs. They even share the same burrow!

Did you know? Your fear of spiders might be written in your DNA. Scientists have discovered that even babies show signs of stress (dilated pupils) when shown pictures of spiders, but not when shown pictures of flowers or fish. It seems humans are born with a "danger" instinct to avoid eight-legged shapes.

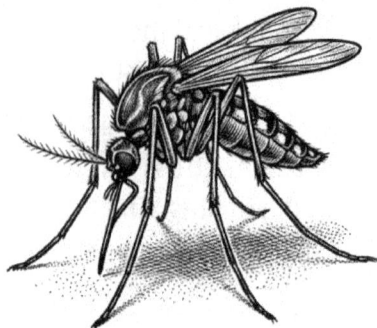

Mosquito
Culicidae

COMMON LOCATION: Worldwide.
A blood-feeding insect linked to disease, itching, and long-standing human fear.

Fire Ant
Solenopsis spp.

COMMON LOCATION: Americas.
Aggressive ants that swarm and sting together, intensifying fear through numbers.

Scorpion
Scorpiones

COMMON LOCATION: Warm regions worldwide.
Ancient arachnids feared for venomous stings and nighttime encounters.

Locust
Acrididae

COMMON LOCATION: Africa, Asia, and beyond.
Swarming insects tied to legends of famine, destruction, and divine punishment.

STORY MOMENT

STORY MOMENT

Story Moment - The Field of Living Diamonds

It is pitch black, and your flashlight carves a narrow cone of light through the tall grass. The night air is cool and still, broken only by the soft crunch of your boots on dry earth. You sweep the beam slowly, searching for the path, when suddenly the ground ahead of you begins to glow.

Scattered through the grass are hundreds of tiny blue-green sparkles, glittering like spilled diamonds. They don't shimmer like dew, and they're far too bright to be stones. The field looks dusted with stars, as if the night sky has fallen to the ground. Curious-and a little uneasy-you kneel and move the light closer to one of the shining points.

The "diamond" shifts.

You freeze. The sparkle isn't a gem at all. You're staring into a face. A wolf spider lifts slightly, and its large central eyes flash as they catch your beam, reflecting it straight back at you like polished mirrors. You straighten quickly and sweep the flashlight across the field again.

The glitter spreads.

Everywhere the beam moves, more lights flare to life-dozens, then hundreds, then far more than you can count. The grass itself seems to watch you. What looked like a quiet field is suddenly alive, filled with unseen hunters standing perfectly still, their eyes glowing only because you are there to reveal them.

You slowly lower the light. The diamonds vanish. The field returns to darkness, calm and ordinary again. But now you know the truth: the night was never empty. It was waiting.

? / QUIZ

FUN QUIZ

1. Why do scientists think humans may be naturally afraid of spiders?
   A) Spiders were once much larger
   B) Early humans learned fear from stories
   C) Quick detection of spiders improved survival
   D) Spiders hunt humans at night

2. True or False: Scorpions glow under ultraviolet light because of chemicals in their exoskeleton.

3. What signal do bed bugs use to know when it's safe to come out and feed at night?
   A) Darkness alone
   B) Body heat
   C) Carbon dioxide from breathing
   D) Sound vibrations

4. Which insect inspired ancient Egyptian beliefs about the sun's daily journey?
   A) Firefly
   B) Scarab beetle
   C) Dragonfly
   D) Locust

5. True or False: Camel spiders chase humans because they are aggressive predators.

6. What makes Wolf Spiders especially frightening when disturbed?
   A) Their venom strength
   B) Their speed
   C) Their ability to jump
   D) Hundreds of babies scattering at o

QUIZ ANSWERS

1. **C**
2. **True**
3. **C**
4. **B**
5. **False**
6. **D**

Mason Bee
Osmia spp.

COMMON LOCATION:
Worldwide.

DESCRIPTION:
A solitary bee that builds nests from mud and plant material, sealing each chamber like a miniature stone mason.

THE CRAWL ZONE™

Pages for the Seriously Curious

ABOUT THE BONUS SECTION

Welcome to **The Crawl Zone**™ - the off-map part of the book where normal rules don't apply. These pages aren't organized like the chapters you just read, and that's on purpose. This is where insects and tiny monsters are pushed to their absolute limits, stripped down to their strangest behaviors, and examined in ways that don't fit neatly anywhere else.

Inside, you'll find record-breaking extremes, blunt answers to questions people are afraid to ask, real behaviors that sound made-up, creatures that only reveal themselves when no one's watching, and mysteries that science still can't fully explain. Nothing here is softened, cleaned up, or wrapped in tidy conclusions.

If the main book made you curious... this section may make you a bit **uneasy**. Walk carefully and watch for those creepy crawlers.

THE CRAWL ZONE - CONTENTS

ZONE 1 – INSECT RECORD BOOK

Some insects don't just survive - they dominate categories scientists didn't even know needed records. The creatures in this file aren't famous because they're common or useful. They're here because they pushed biology past what feels reasonable, sometimes past what feels safe. **These are extremes - verified, measured, and unsettlingly real.** (note: records change over time, but as of this date these are the real facts)

The heaviest insect ever recorded is the **giant wētā** (**New Zealand**, large flightless cricket, nocturnal omnivore). Some individuals weigh more than a small bird, with thick armored legs built for climbing, fighting, and surviving cold mountain nights. Holding one feels less like touching a bug and more like handling a living stone with joints.

The longest insect on Earth is the **stick insect Phryganistria chinensis** (**China**, leaf-mimicking insect, camouflage specialist). Fully stretched, it can exceed 24 inches from end to end, longer than many human arms. In the wild, most people walk past them without ever realizing they just passed an animal longer than their forearm.

The fastest flying insect is the **horsefly** (**worldwide**, blood-feeding fly, ambush hunter). Clocked at speeds approaching 90 miles per hour, it can outrun most birds over short bursts. You don't feel hunted - until it lands.

The loudest insect on the planet is the **male cicada** (**worldwide**, sap-feeding insect, sound-based mating caller). At close range, its call can exceed 120 decibels, louder than a chainsaw. The sound isn't accidental - it's pressure-built vibration amplified by hollow body chambers designed like acoustic weapons.

The strongest insect relative to size is the **dung beetle** (**worldwide**, decomposer beetle, nutrient recycler). Some species can pull over 1,000 times their own body weight. Scaled to human size, that would be equivalent to dragging multiple loaded freight trains without mechanical aid.

The most venomous insect sting belongs to the **bullet ant** (**Central and South America**, predatory ant, territorial defender). Victims describe the pain as pure, electrical agony that can last more than 24 hours. There is no antidote - only time, and regret.

The most heat-resistant insect is the **desert ant** (**Sahara and Middle East**, scavenger ant, extreme-temperature runner). It forages in surface temperatures hot enough to kill most animals in minutes. Its legs are elongated to lift its body off the ground, and its navigation system is so precise it can sprint home before its brain overheats.

The coldest-surviving insect is the **Antarctic midge** (**Antarctica**, wingless fly, freeze-tolerant survivor). It can survive being frozen solid for most of the year. When thawed, it resumes movement as if nothing happened.

These records aren't here to impress you.
They exist to remind you that evolution doesn't care what feels comfortable, familiar, or safe. Somewhere nearby - underground, in the walls, in the trees, or in the dark - something smaller than your thumb may already hold a record that hasn't been measured yet.

ZONE 2 - AFRAID TO ASK?

QUESTIONS MANY PEOPLE ARE AFRAID TO GOOGLE

Some questions don't sound childish - they sound dangerous. Not because the answers are violent, but because they strip away the comfort of "probably not" and replace it with "actually... yes." The questions in this file are real, researched, and asked quietly by people who weren't sure they wanted to know the answers.

Can insects see me in the dark?

Yes - many can. **Moths** and **cockroaches** (**worldwide**, nocturnal insects, low-light navigators) use compound eyes designed to gather and stack tiny amounts of light, allowing them to detect motion long before human eyes adjust. Darkness protects you far less than stillness does.

Do bugs feel pain?

They don't feel pain the way humans do, but that doesn't mean they feel nothing. Insects possess **nociceptors**, sensory systems that detect damage and trigger avoidance behaviors. When an insect struggles, withdraws, or self-amputates a limb, it's responding to injury - not emotion, but awareness.

Can insects crawl into my ear while I sleep?

It's rare, but documented. **Cockroaches**, **ants**, and **beetles** (**worldwide**, opportunistic insects, shelter seekers) are attracted to warmth, moisture, and stillness. Most incidents end quickly, but removal often requires medical assistance - not because of danger, but because of panic.

Are there insects living in my walls right now?

Possibly. **Termites**, **carpenter ants**, and **powderpost beetles** (**worldwide**, wood-associated insects, structural colonizers) can live undetected for years. Silence doesn't mean absence - it often means the colony is thriving.

Can insects recognize individual humans?
Some can. **Wasps** and **bees** (**worldwide**, social insects, visual learners) have been shown to recognize human faces and remember threats. If one individual is targeted, it's not random - it's learned behavior.

Do insects ever watch instead of attack?
Yes. **Praying mantises** and **jumping spiders** (**worldwide**, visual predators, ambush hunters) will remain motionless for long periods while tracking prey with precise head movements. What feels like imagination is often observation.

Can insects survive inside the human body?
Certain species can - temporarily. **Botflies** and **screwworm larvae** (**Americas**, parasitic flies, tissue feeders) require living hosts during development. Infection isn't common, but it is real, treatable, and deeply unpleasant.

Why do bugs freeze when I look at them?
Many insects rely on motion to avoid detection. **Cockroaches** and **stick insects** (**worldwide**, survival strategists, motion-based evaders) stop instantly when light or vibration changes. The pause isn't confusion - it's calculation.

These answers aren't meant to scare you.
They're meant to replace vague fear with clear understanding - because what we imagine is often worse than what's actually happening. Still, once you know how much insects can see, feel, remember, and endure... the quiet starts to feel different.

ZONE 3 - REAL BEHAVIORS

REAL INSECT BEHAVIORS THAT SOUND ILLEGAL
Verified by Science. Approved by Evolution.

Some behaviors are so extreme that they sound like mistakes or crimes - against biology itself. But insects don't follow human rules. They follow survival math. The creatures in this file hijack minds, rewrite bodies, abandon organs, and weaponize other living things in ways that feel wrong... yet work perfectly.

These aren't myths. These are documented bug strategies.

The ant that loses its mind on purpose is the **zombie ant** (**tropical regions**, parasitized ant, host for mind-controlling fungus). Infected ants abandon their colony, climb vegetation, and clamp down before dying exactly where the fungus needs them. The parasite then erupts from the ant's head, releasing spores onto new victims below.

The wasp that rewrites behavior is the **emerald cockroach wasp** (**Africa**, parasitic wasp, neural manipulator). It injects venom directly into a cockroach's brain, removing free will without killing the host. The cockroach walks calmly into its own burial chamber, alive, where it becomes food.

The insect that dissolves itself is the **mayfly** (**worldwide**, short-lived aquatic insect, reproductive specialist). Adults lack functional mouths and digestive systems because they don't need them. Their sole purpose is to reproduce - sometimes within hours - then die.

The caterpillar that hires bodyguards is the **lycaenid butterfly larva** (**worldwide**, mutualistic caterpillar, chemical communicator). It secretes sugary fluids that attract ants, which defend it violently in exchange. The caterpillar survives not by hiding, but by outsourcing protection.

The beetle that explodes on command is the **bombardier beetle** (**worldwide**, ground beetle, chemical defender). When threatened, it mixes volatile chemicals inside its abdomen and fires boiling spray outward with explosive force. The reaction is controlled, repeatable, and precise.

The insect that amputates itself is the **stick insect** (**worldwide**, camouflage insect, limb-regenerating survivor). When trapped, it can deliberately shed a leg to escape. In juveniles, the limb grows back - proof that sacrifice can be strategic.

The fly that eats its host from the inside is the **screwworm larva** (**Americas**, parasitic fly, tissue feeder). Unlike scavengers, it feeds on living flesh, burrowing deeper over time. Untreated infestations can be fatal - not because of toxins, but because of persistence.

The ant colony that wages war is the **army ant swarm** (**tropical regions**, blind predatory ant, collective hunter). Millions move as a single organism, overrunning anything unable to flee. There are no leaders - the behavior emerges from rules followed blindly and perfectly.

The insect that turns enemies into zombies is the **parasitoid wasp** (**worldwide**, parasitic wasp, biological controller). It lays eggs inside living hosts that continue functioning until the larvae mature. Death comes only after usefulness ends.

Nothing here is accidental.
These behaviors evolved because they worked - again and again - over millions of years. When survival demands efficiency, nature doesn't hesitate, compromise, or apologize. It invents solutions that feel uncomfortable only because they succeed so completely.

.

ZONE 4 - WHEN NOT LOOKING

THINGS THAT COME OUT ONLY WHEN YOU'RE NOT LOOKING

The Hidden Schedules of Tiny Hunters

Not all insects hide because they're afraid. Many hide because they've learned exactly when the world belongs to them. Light, sound, vibration, temperature, and even human routines shape invisible schedules that play out every night. While people sleep, clean, and turn off lights, entire ecosystems quietly switch shifts.

This file is about timing - the moments when insects emerge, move, and disappear without being seen.

The insects that wait for silence include **cockroaches** (**worldwide**, nocturnal scavengers, vibration-sensitive foragers). They don't simply avoid light - they monitor sound and pressure changes. A room can look empty for hours, then fill with movement the moment footsteps fade.

The hunters that track shadows are **praying mantises** (**worldwide**, ambush predators, visual trackers). They remain motionless for long periods, rotating their heads to follow movement without shifting their bodies. Many strikes occur only after prey believes it has not been noticed.

The insects that follow human schedules include **bed bugs** (**worldwide**, blood-feeding insects, heat and carbon-dioxide trackers). They emerge during the deepest stages of sleep, guided by breath and body warmth. Their timing is so precise that people often never see them awake.

The night flyers that map the dark are **moths** (**worldwide**, nocturnal insects, low-light navigators). Their eyes are built to gather minimal light, allowing flight under starlight alone. Artificial lights disrupt this system, trapping them in endless spirals that mimic navigation errors.

The ground hunters that hear footsteps are **wolf spiders** (**worldwide**, wandering predators, vibration detectors). They don't rely on webs. Instead, they sense movement through the ground, freezing or advancing based on pressure waves you never notice making.

The insects that appear after rain include **termites** and **flying ants** (**worldwide**, colony builders, synchronized reproducers). Entire populations emerge simultaneously for short mating flights, then vanish again within hours. The event is massive - and almost always missed.

The creatures that avoid eyes, not light are **silverfish** (**worldwide**, moisture-loving insects, fast evaders). They will cross bright rooms if unobserved, but scatter instantly when watched. For them, visibility is measured by attention, not illumination.

The insects that patrol at body temperature are **mosquitoes** (**worldwide**, blood-feeding insects, thermal trackers). Many species become active only when ambient temperatures drop enough to highlight warm-blooded targets. The night sharpens their aim.

Night doesn't just change what you see.
It changes who owns the space. When lights go out and movement slows, a different set of rules takes over - written not in sound or color, but in timing. Most of these creatures aren't hiding from you. They're simply waiting for their turn.

BONUS 5 - UNSOLVED MYSTERIES

INSECT MYSTERIES WITH NO ENDING (YET)

Observed by Science. Still Unfinished.

Some insect patterns don't break the rules - they expose the gaps in them. Scientists can track these events, measure parts of them, and rule out simple causes, yet the full explanation never quite locks in. The closer we study these mysteries, the more they resist tidy answers.

These aren't legends. These are open cases.

Entire bee colonies vanish without a trace.
In events known as **Colony Collapse Disorder** (worldwide, honey bee colonies, sudden worker loss), healthy hives are found nearly empty - the queen, young bees, and food remain, but most adult workers are gone. There are no piles of bodies, no obvious poison, no smoking gun. Something pushes the system past a tipping point... and then it breaks.

Flying insects are quietly disappearing in some places.
Long-term monitoring has revealed dramatic drops in flying insect populations (Europe and beyond, mixed insect species, ecosystem indicators). In certain protected areas, the air itself has become emptier over time. The mystery isn't just why - it's how changes can happen so fast without anyone noticing until silence sets in.

Thousands of fireflies suddenly blink as one.
In rare displays, **synchronous fireflies** (Appalachian Mountains and Asia, flashing beetles, group signalers) light up forests in perfect unison. No leader gives the signal. No master clock tells them when to start. Somehow, individual insects merge into a single, pulsing system - and scientists are still arguing over how timing locks in so precisely.

Some insects keep time underground for decades.
Periodical cicadas (eastern United States, long-lived insects, mass emergers) spend 13 or 17 years hidden below ground, then emerge together in numbers so large they overwhelm predators. How an insect with no calendar, no sunlight, and no reference points counts years so accurately remains one of nature's strangest timekeeping feats.

Migrating insects know where they're going - without ever having been there.
Monarch butterflies (**North America**, long-distance migrants, multi-generation travelers) complete journeys spanning thousands of miles, even though no single butterfly makes the full round trip. Direction comes from the sun, internal clocks, and possibly Earth's magnetic field - but how these signals combine into a reliable mental map is still not fully understood.

ZONE 6 - THE RESEARCH VAULT

THE RESEARCH VAULT

For Readers Who Want the Receipts

Straight up truth: this Zone isn't for everyone. The resources here go deeper - way deeper - into the technical science behind the mysteries you just read. These links lead to real studies, long papers, and serious research written by scientists for scientists.

You don't need to read any of it to enjoy this book. But if you've ever wondered *"Is this actually real?"* or *"How do they know that?"* - this is where the evidence lives. **MIND-BLOWN™**

The **"vanishing hive" problem** is known as **Colony Collapse Disorder** (**worldwide**, honey bee crisis pattern, worker-loss collapse). The signature is eerie: most **worker bees** disappear, leaving behind a **queen**, **brood**, and **food stores**-often with few dead bees nearby. The scary part isn't that we know nothing; it's that the evidence keeps pointing to *multiple* pressures stacking at once, and the mix can change by region and year. **QR Code Link: Environmental Protection Agency**

The **insect decline mystery** is bigger than one species: long-term monitoring has documented steep drops in **flying insect biomass** in protected areas in Germany over decades, sparking a global debate about how widespread declines are and what's driving them. Even with newer follow-up work and better methods, the trend question isn't "solved"-it's a moving target that depends on habitat, region, climate shifts, and how we measure insects in the first place. **QR Code Link: PLOS Research**

PLOS is a non-profit organization on a mission to drive open science forward with measurable, meaningful change in research publishing, policy, and practice.

The fireflies that invent a rhythm are **synchronous fireflies** (**U.S. Appalachians and parts of Asia**, flashing beetles, swarm signalers). Individually, some species don't flash with a steady internal beat-yet in large groups they suddenly lock into coordinated bursts, like a living light machine. Researchers can model pieces of this behavior, but the full "how do thousands become one clock?" effect still feels like emergent magic you can't fully reduce to a single simple rule. **QR Code Link: eLife**

eLife is a non-profit organization inspired by research funders and led by scientists. Our mission is to help scientists accelerate discovery by operating a platform for research communication that encourages and recognizes the most responsible behaviors in science.

The monarch migration "how" is still not fully closed: **monarch butterflies** (**North America**, long-distance migrants, multi-cue navigators) use a **time-compensated sun compass** linked to circadian timing, and evidence suggests magnetic sensing can contribute under certain conditions. What remains tricky is exactly **when** each cue dominates in the real world, how they integrate cues under messy weather/light conditions, and how that navigation holds across generations. **QR Code Link: NLM**

NLM: National Library of Medicine. National Center for Biotechnology Information advances science and health by providing access to biomedical and genomic information.

These mysteries end the book for a reason. Because the most unsettling truth isn't "bugs are scary."
It's this:

Even now-after microscopes, genetics, radar tracking, and decades of field work-**the small world still keeps secrets.**

BUG RESOURCES w/ QR Codes

Trusted Sites for Curious Minds (Kids & Adults)

Not all science lives behind paywalls or college textbooks. These sites explore insects, tiny creatures, and the hidden world around us in ways that are visual, fascinating, and easy to understand - whether you're a curious kid, a parent, or a lifelong learner.

🐜 BIG-PICTURE BUG SCIENCE (Friendly & Visual)

National Geographic – Insects & Bugs
Stunning photography, clear explanations, and real science written for general readers.
https://www.nationalgeographic.com/animals/invertebrates/

Smithsonian National Museum of Natural History
Reliable, well-written resources on insects, ecosystems, and evolution.
https://naturalhistory.si.edu

BBC Earth – Insects & Minibeasts
Short articles and videos that explain strange behaviors in simple terms.
https://www.bbcearth.com

🪲 BUGS UP CLOSE (KID-FRIENDLY EXPLORATION)

BugGuide.net
A photo-based guide to insects of North America, great for identification and curiosity.
https://bugguide.net

iNaturalist
Explore real insect sightings from around the world - and add your own.
https://www.inaturalist.org

National Wildlife Federation – Insects
Easy explanations focused on nature and conservation.
https://www.nwf.org

🕷 CREEPY, WEIRD & AMAZING (STILL TRUSTED)

Australian Museum – Insects & Arachnids
Fantastic balance of creepy and educational, with clear explanations.
https://australiamuseum.net.au

American Museum of Natural History – Arthropods
Deep enough for adults, readable enough for teens.
https://www.amnh.org

California Academy of Sciences – Insect Science
Hands-on science, videos, and real-world research explained simply.
https://www.calacademy.org

🌐 NATURE, ECOSYSTEMS & WHY INSECTS MATTER

Xerces Society for Invertebrate Conservation
Explains why insects matter - without getting preachy or technical.
https://www.xerces.org

EPA – Pollinators & Insects
Clear, factual resources on bees, insects, and ecosystems.
https://www.epa.gov/pollinator-protection

National Park Service – Insects
Shows how insects fit into real landscapes kids recognize.
https://www.nps.gov

Ask A Biologist (Arizona State University)
Real scientists answering real questions in kid-friendly language.
https://askabiologist.asu.edu

🔬 FOR READERS WHO WANT A LITTLE MORE DEPTH (OPTIONAL - Adult supervision recommended)

SciShow (YouTube)
Short, energetic videos that explain complex insect science clearly.
https://www.youtube.com/@SciShow

Kurzgesagt – In a Nutshell
Big science ideas, beautifully explained (select insect-related topics).
https://www.youtube.com/@kurzgesagt

Act Now - offer may be gone before the next molt

GET FREE eBOOK

Sign up at our secure website

https://MindBlownBooks.com

Paperbacks, e-Books, Activity Books,

Audio Books + More

THANK YOU FOR EXPLORING WITH US

We're grateful you spent your time with these stories, facts, myths, and legends... and we hope at least one of them made you stop and say, *"No way... that can't be real."*

Curiosity is powerful. When you follow it, the world of any size opens up in ways you never expected. You showed up ready to explore. Awesome!

**Thanks again for reading & wondering.
And for being the kind of person who looks up.**

See you in the next book. **Coming Soon.**

The Official Stamp of Awesome Weirdness

www.ingramcontent.com/pod-product-compliance
Lightning Source LLC
Chambersburg PA
CBHW050113280326
41933CB00010B/1085